INDUCTION, SYNCHRONOUS AND SPECIAL ELECTRICAL MACHINES

AUTHOR (S)

Dr. K. Selvajyothi
Associate Professor,
IIITDM Kancheepuram,
Chennai

Dr. S. Tamilselvi
Associate Professor,
Sri Sivasubramaniya Nadar
College of Engineering,
Chennai

Prof. Sarat Kumar Sahoo
FIE, C. Engg., SMIEEE,
Parala Maharaja Engineering
College, Odisha

REVIEWER

Dr. Nikunj Patel
Assistant Professor,
Gujarat Technological University (GTU)

KHANNA BOOK PUBLISHING CO. (P) Ltd.
4C/4344, Ansari Road, Darya Ganj,
New Delhi, India

All India Council for Technical Education
Nelson Mandela Marg, Vasant Kunj,
New Delhi, India

ISBN: 978-93-5538-613-7

Book Code: DIP475EN

Induction, Syncronous And Special Electrical Machines

by Dr. K. Selvajyothi, Dr. S. Tamilselvi and Prof. Sarat Kumar Sahoo

First Edition: 2026

Originally Published by:
All India Council for Technical Education, Ministry of Education, GOI

Exclusively Published & Distributed by:
Khanna Book Publishing Co. (P) Ltd.
4C/4344, Ansari Road, Daryaganj,
New Delhi (INDIA)

Visit us at: **www.khannabooks.com**
Write us at: **contact@khannabooks.com**

To view complete list of books, Please scan the QR Code: KPH

Printed in Bharat.

प्रो. टी. जी. सीताराम
अध्यक्ष
Prof. T. G. Sitharam
Chairman

सत्यमेव जयते

आज़ादी का अमृत महोत्सव

अखिल भारतीय तकनीकी शिक्षा परिषद्
(भारत सरकार का एक सांविधिक निकाय)
(शिक्षा मंत्रालय, भारत सरकार)
नेल्सन मंडेला मार्ग, वसंत कुंज, नई दिल्ली-110070
दूरभाष : 011-26131498
ई-मेल : chairman@aicte-india.org

ALL INDIA COUNCIL FOR TECHNICAL EDUCATION
(A STATUTORY BODY OF THE GOVT. OF INDIA)
(Ministry of Education, Govt. of India)
Nelson Mandela Marg, Vasant Kunj, New Delhi-110070
Phone : 011-26131498
E-mail : chairman@aicte-india.org

FOREWORD

Engineers are the backbone of any modern society. They are the ones responsible for the marvels as well as the improved quality of life across the world. Engineers have driven humanity towards greater heights in a more evolved and unprecedented manner.

The All India Council for Technical Education (AICTE), have spared no efforts towards the strengthening of the technical education in the country. AICTE is always committed towards promoting quality Technical Education to make India a modern developed nation emphasizing on the overall welfare of mankind.

An array of initiatives has been taken by AICTE in last decade which have been accelerated now by the National Education Policy (NEP) 2020. The implementation of NEP under the visionary leadership of Hon'ble Prime Minister of India envisages the provision for education in regional languages to all, thereby ensuring that every graduate becomes competent enough and is in a position to contribute towards the national growth and development through innovation & entrepreneurship.

One of the spheres where AICTE had been relentlessly working since past couple of years is providing high quality original technical contents at Under Graduate & Diploma level prepared and translated by eminent educators in various Indian languages to its aspirants. For students pursuing 2nd year of their Engineering education, AICTE has identified 87 books, which shall be translated into 12 Indian languages - Hindi, Tamil, Gujarati, Odia, Bengali, Kannada, Urdu, Punjabi, Telugu, Marathi, Assamese & Malayalam. In addition to the English medium, books in different Indian Languages are going to support the students to understand the concepts in their respective mother tongue.

On behalf of AICTE, I express sincere gratitude to all distinguished authors, reviewers and translators from the renowned institutions of high repute for their admirable contribution in a record span of time.

AICTE is confident that these outcomes based original contents shall help aspirants to master the subject with comprehension and greater ease.

(Prof. T. G. Sitharam)

ACKNOWLEDGEMENT

The authors are grateful to the authorities of AICTE, particularly Prof. (Dr.) T. G. Sitharam, Chairman; Dr. Abhay Jere, Vice-Chairman, Prof. Rajive Kumar, Member-Secretary, Dr. Ramesh Unnikrishnan, Advisor-II and Dr. Sunil Luthra, Director, Training and Learning Bureau for their planning to publish the books on ***Induction, Synchronous and Special Electrical Machines***. We sincerely acknowledge the valuable contributions of the reviewer of the book Dr Nikunj patel, Assistant Professor, Electrical Engineering (EE), Gujarat Technological University (GTU), Gujarat for making it students' friendly and giving a better shape in an artistic manner. We would like to thank our research scholars helping in reading and editing the contents.

This book is an outcome of various suggestions of AICTE members, experts and authors who shared their opinion and thought to further develop the engineering education in our country. Acknowledgements are due to the contributors and different workers in this field whose published books, review articles, papers, photographs, footnotes, references and other valuable information enriched us at the time of writing the book.

Dr. K. Selvajyothi

Dr. S. Tamilselvi

Prof. Sarat Kumar Sahoo

PREFACE

Electrical machines are the backbone of modern industrial and technological advancements. The power our homes, drive our factories, and enable the devices that have become integral to our daily lives. Understanding the principles and applications of electrical machines is not only a fundamental aspect of electrical engineering but is also essential for anyone seeking a career in the electrical and electronics industry.

This book, "Induction, Synchronous, and Special Electrical Machines" is crafted with a primary focus on the comprehensive understanding of electrical machines, catering to the needs of diploma students in electrical and electronics engineering. Whether you are a student embarking on your academic journey, an instructor guiding aspiring minds, or a practitioner in the field, this book aims to be your reliable companion.

Key Features of the Book:

Comprehensive Coverage: *The book covers a wide spectrum of electrical machines, including induction motors, synchronous machines, and specialized machines such as stepper motors, servomotors, and single-phase motors.*

Accessible Language: *We have strived to present complex concepts in an easily understandable language, ensuring that students at the diploma level can grasp the fundamentals and applications of electrical machines with confidence.*

Practical Insights: *Real-world examples and case studies are integrated throughout the book to bridge the gap between theoretical knowledge and practical application. These insights will help students appreciate the relevance and significance of the subject matter.*

Illustrative Visuals: *Richly illustrated diagrams, schematics, and phasor diagrams accompany the text, enhancing the reader's ability to visualize and comprehend the intricate workings of electrical machines.*

Exercises and Problem Solving: *Each chapter includes exercises and problems that encourage active learning and application of the concepts. Solutions to selected problems are provided to assist students in their self-assessment and progress.*

Emerging Trends: *Recognizing the ever-evolving nature of the field, the book also touches upon emerging trends in electrical machines, such as the application of power electronics and renewable energy integration.*

Target Audience: *This book is tailored to meet the needs of diploma students pursuing electrical and electronics engineering. It serves as an ideal resource for diploma-level courses in electrical machines, offering a strong foundation for those who wish to pursue higher studies or enter the workforce directly.*

Acknowledgments: *We would like to express our gratitude to the educators, students, and professionals who have contributed to the development of this book through their insights, expertise, and valuable feedback. Their dedication to the field of electrical machines has been instrumental in creating this comprehensive resource.*

We invite all readers to embark on a journey through the pages of "Induction, Synchronous, and Special Electrical Machines for Diploma Students." May this book serve as a guiding light, illuminating the path to a deeper understanding of electrical machines and inspiring the engineers and innovators of tomorrow.

Dr. K. Selvajyothi
Dr. S. Tamilselvi
Prof. Sarat Kumar Sahoo

OUTCOME BASED EDUCATION

For the implementation of an outcome based education the first requirement is to develop an outcome based curriculum and incorporate an outcome based assessment in the education system. By going through outcome based assessments, evaluators will be able to evaluate whether the students have achieved the outlined standard, specific and measurable outcomes. With the proper incorporation of outcome based education there will be a definite commitment to achieve a minimum standard for all learners without giving up at any level. At the end of the programme running with the aid of outcome based education, a student will be able to arrive at the following outcomes:

Programme Outcomes (POs) are statements that describe what students are expected to know and be able to do upon graduating from the program. These relate to the skills, knowledge, analytical ability attitude and behaviour that students acquire through the program. The POs essentially indicate what the students can do from subject-wise knowledge acquired by them during the program. As such, POs define the professional profile of an engineering diploma graduate.

National Board of Accreditation (NBA) has defined the following seven POs for an Engineering diploma graduate:

PO1. Basic and Discipline specific knowledge: Apply knowledge of basic mathematics, science and engineering fundamentals and engineering specialization to solve the engineering problems.

PO2. Problem analysis: Identify and analyses well-defined engineering problems using codified standard methods.

PO3. Design/ development of solutions: Design solutions for well-defined technical problems and assist with the design of systems components or processes to meet specified needs.

PO4. Engineering Tools, Experimentation and Testing: Apply modern engineering tools and appropriate technique to conduct standard tests and measurements.

PO5. Engineering practices for society, sustainability and environment: Apply appropriate technology in context of society, sustainability, environment and ethical practices.

PO6. Project Management: Use engineering management principles individually, as a team member or a leader to manage projects and effectively communicate about well-defined engineering activities.

PO7. Life-long learning: Ability to analyse individual needs and engage in updating in the context of technological changes.

COURSE OUTCOMES

By the end of the course the students are expected to learn:

CO-1: Gain comprehensive understanding of AC and Special electrical machines, including their working principles, constructional details, and torque-speed characteristics.

CO-2: Develop proficiency in analysing rotor quantities, such as frequency, induced EMF, and power factor, and their impact on motor performance. Understanding the power flow diagram and operation in all four quadrants.

CO-3: Acquire knowledge of starters, speed control methods, and motor selection criteria for different applications, ensuring efficient operation and maintenance of AC motors.

CO-4: Explore the principles of operation of three-phase alternators and synchronous motors, including torque analysis, voltage regulation, and maintenance practices.

CO-5: Understand the diverse applications of special AC machines, including synchronous reluctance motors, switched reluctance motors, BLDC motors, and stepper motors.

Mapping of Course Outcomes with Programme Outcomes to be done according to the matrix given below:

Course Outcomes	Expected Mapping with Programme Outcomes (1- Weak Correlation; 2- Medium correlation; 3- Strong Correlation)						
	PO-1	**PO-2**	**PO-3**	**PO-4**	**PO-5**	**PO-6**	**PO-7**
CO-1	3	3	3	3	1	2	3
CO-2	3	3	3	3	1	2	3
CO-3	3	3	3	3	1	2	3
CO-4	3	3	3	3	1	2	3
CO-5	3	3	3	3	1	2	3

GUIDELINES FOR TEACHERS

To implement Outcome Based Education (OBE) knowledge level and skill set of the students should be enhanced. Teachers should take a major responsibility for the proper implementation of OBE. Some of the responsibilities (not limited to) for the teachers in OBE system may be as follows:

- Within reasonable constraint, they should manoeuvre time to the best advantage of all students.
- They should assess the students only upon certain defined criterion without considering any other potential ineligibility to discriminate them.
- They should try to grow the learning abilities of the students to a certain level before they leave the institute.
- They should try to ensure that all the students are equipped with the quality knowledge as well as competence after they finish their education.
- They should always encourage the students to develop their ultimate performance capabilities.
- They should facilitate and encourage group work and team work to consolidate newer approach.
- They should follow Blooms taxonomy in every part of the assessment.

Bloom's Taxonomy

Level	Teacher should Check	Student should be able to	Possible Mode of Assessment
Create	Students ability to create	Design or Create	Mini project
Evaluate	Students ability to justify	Argue or Defend	Assignment
Analyse	Students ability to distinguish	Differentiate or Distinguish	Project/Lab Methodology
Apply	Students ability to use information	Operate or Demonstrate	Technical Presentation/ Demonstration
Understand	Students ability to explain the ideas	Explain or Classify	Presentation/Seminar
Remember	Students ability to recall (or remember)	Define or Recall	Quiz

GUIDELINES FOR STUDENTS

Students should take equal responsibility for implementing the OBE. Some of the responsibilities (not limited to) for the students in OBE system are as follows:

- Students should be well aware of each UO before the start of a unit in each and every course.
- Students should be well aware of each CO before the start of the course.
- Students should be well aware of each PO before the start of the programme.
- Students should think critically and reasonably with proper reflection and action.
- Learning of the students should be connected and integrated with practical and real life consequences.
- Students should be well aware of their competency at every level of OBE.

ABBREVIATIONS AND SYMBOLS

List of Abbreviations

General Terms			
Abbreviations	**Full form**	**Abbreviations**	**Full form**
AC	Alternating Current	**MMF**	Magneto Motive Force
DC	Direct Current	**PPM**	Pulse Position Modulation
GE	General Electric	**LIM**	Linear Induction Motor
HP	Horse Power	**RL LOAD**	Inductive Load
ABB	Asea Brown Boveri	**RC LOAD**	Capacitive Load
EMF	Electromotive Force	**ZPF**	Zero Power Factor
RMS	Root Mean Square	**OCC**	Open Circuit Characteristic
IM	Induction Motor	**SCC**	Short Circuit Characteristic
VFD	Variable Frequency Drives	**OC**	Open Circuit
PF	Power Factor	**SC**	Short Circuit
RPM	Revolutions Per Minute	**TPSTS**	Triple Pole Single Throw Switch
MAX	Maximum	**DPSTS**	Double Pole Single Throw Switch
VA	Volt-Ampere	**MI**	Moving Iron
KVAR	Kilo Volt Ampere Reactive	**MC**	Moving Coil
HVAC	Heating, Venting, And Air Conditioning	**ASA**	American Standard Association
DOL	Direct Online Starter	**KVL**	Kirchhoff's Voltage Law
SCR	Silicon Controlled Rectifiers	**BHP**	Brake Horsepower
VVVF	Variable Voltage Variable Frequency	**BLDC**	Brushless Direct Current
VSD	Variable Speed Drive	**FHP**	Fractional Horse Power
MOSFET	Metal Oxide Semiconductor Field Effect Transistor	**PMBLDC**	Permanent Magnet Brushless Direct Current
IGBT	Insulated Gate Bipolar Transistor	**PMSM**	Permanent Magnet Synchronous Motor
PWM	Pulse Width Modulation	**BLPM**	Brushless Permanent Magnet
PSC	Permanent Split Capacitor	**VR**	Variable Reluctance
RMF	Rotating Magnetic Field	**SRM**	Switched Reluctance Motor
CS	Centrifugal Switch	**CNC**	Computerised Numerical Control

General Terms			
Abbreviations	**Full form**	**Abbreviations**	**Full form**
HIPOT	High Potential Testing	**UPF**	Unity Power Factor

List of Symbols

Symbols	Description	Symbols	Description
ϕ	Magnetic Flux	V_1	Stator Voltage
N_s	Main Flux Speed	V_2	Rotor Voltage
N_r	Rotor Speed	δ	Torque Angle
s	Slip	Z	Impedance
I_r	Rotor Current	P_o	Output Power
p	No of Poles	P_{rc}	Rotor Copper Losses
Hz	Hertz	P_{fw}	Friction And Windage Losses
e_r	Induced Emf in The Rotor	P	Active Power
e_s	Induced Emf in The Stator	Q	Reactive Power
K_{ws}	Stator Winding Factor	S	Apparent Power
K_{wr}	Rotor Winding Factor	P_{loss}	Power Losses
%	Percentage	η	Efficiency
3- Ø	Three Phase	F	Farad
τ	Torque	K_d	Distribution Factor
τ_{max}	Maximum Torque	Z_S	Synchronous Impedance
τ_{stg}	Starting Torque	X_L	Leakage Reactance
FL	Full Load	X_S	Synchronous Reactance
N_{FL}	Speed At Full Load	I_a	Armature Current
T_{FL}	Full Load Torque	R_a	Armature Resistance
T_{MAX}	Maximum Torque	V_t	Terminal Voltage
V_L	Line Voltage	$I_a R_a$	Armature Resistance Voltage Drop
I_L	Line Current	$I_a X_s$	Reactive Voltage Drop
I_{SC}	Short Circuit Current	S_i	Complex Power Input
P_M	Mechanical Power	P_i	Real Input Power
f	Frequency	Q_i	Reactive Input Power
q	Angle	S_o	Complex Power Output
K_p	Pitch Factor	P_o	Real Output Power
Ω	Ohm	Q_o	Reactive Output Power
r_1	Stator Resistance	$P_{suffix\ max}$	Maximum Power Developed

Symbols	Description	Symbols	Description
X_1	Stator Leakage Reactance	P_{hys}	Hysteresis Losses
r_2	Rotor Resistance	B_{max}	Maximum Flux Density
X_2	Rotor Leakage Reactance	P_{eddy}	Eddy Current Losses
I_o	No-Load Current	E_b	Back Emf
I_1	Stator Current	α	Load Angle
I_2	Rotor Current	$\circ$	Degree

LIST OF FIGURES

Unit 1 Three-Phase Induction Motor – Working and Construction

Unit 2 Three-Phase Induction Motor – T-N Characteristics

Unit 5 Three-phase Alternator – Working and Construction

Unit 6 Three-phase Alternator –Voltage Regulation

Unit 7 Synchronous Motor

Unit 8 Fractional Horsepower (FHP) Motors – Working and Construction

CONTENTS

Unit 2: Three-Phase Induction Motor – T-N Characteristics **23-54**

Unit 3: Three-Phase Induction Motor - Starting and Speed Control **55-82**

1 THREE-PHASE INDUCTION MOTOR- Working and Construction

UNIT SPECIFICS

Through this module we have discussed the following aspects:

- *Familiarizing the working of Induction motor;*
- *Concept of rotating magnetic field and its speed;*
- *Concept of rotor speed and slip;*
- *Constructional details of 3-phase Induction motor;*
- *Types of 3-phase Induction Motors;*
- *Frequency of rotor emf and current;*
- *Problems.*

The photos of practical motor are shown to detail the various parts of the machine. The real time applications of the topics are discussed for inducing further curiosity and creativity.

Besides giving a large number of multiple-choice questions as well as questions of short and long answer types marked in two categories following lower and higher order of Bloom's taxonomy, assignments through a number of numerical problems, a list of references and suggested readings are given in the unit so that one can go through them for practice. It is important to note that for getting more information on various topics of interest some QR codes have been provided in different sections which can be scanned for relevant supportive knowledge.

After the related practical, based on the content, there is a "Know More" section. This section has been carefully designed so that the supplementary information provided in this part becomes beneficial for the users of the book. This section mainly highlights the initial activity, examples of some interesting facts, analogy, history of the development of the subject focusing the salient observations and finding, timelines starting from the development of the concerned topics up to the recent time, applications of the subject matter for our day-to-day real life or/and industrial applications on variety of aspects, case study related to environmental, sustainability, social and ethical issues whichever applicable, and finally inquisitiveness and curiosity topics of the unit.

RATIONALE

Induction motor is the work horse of any industry. It has a widespread application in industrial and domestic applications. About 50% of the power consumption is due to the induction motor loads. This chapter introduces the students to the construction and working principle of the motor. To understand the fundamentals in the principle of operation very important aspect is the self-starting capability of three phase induction motor due to revolving field.

PRE-REQUISITES

Machines I: Sem III

Physics: Electromagnetism (Class X)

UNIT OUTCOMES

List of outcomes of this unit is as follows:

U1-O1: Understand construction of the motor

U1-O2: Able to understand the constructional difference in squirrel cage and slip ring induction motor

U1-O3: Feel the production of rotating flux in the stator

U1-O4: How self-starting is possible in induction motor

U1-O5: Able to get to know about the significance of slip

Unit-1 Outcomes	EXPECTEDMAPPINGWITHCOURSEOUTCOMES (1-WeakCorrelation;2-Mediumcorrelation;3-StrongCorrelation)				
	CO-1	CO-2	CO-3	CO-4	CO-5
U1-O1	3	-	-	-	-
U1-O2	3	-	-	-	-
U1-O3	3	-	-	-	-
U1-O4	3	-	-	-	-
U1-O5	-	3	-	-	-

1.1 INTRODUCTION

The main energy source for the world is fossil fuel-based power plants. The invention of steam engines during industrial revolution helped in conserving thermal energy into electrical energy. This electrical energy is available as alternating current (AC) source. British scientist Michael Faraday invented (1821) that by placing a current carrying conductor in a magnetic field the conductor experiences a torque due to the interaction of electrical current and magnetic field. Thus, the electrical energy is converted to mechanical energy. This has paved the way to design the most primitive of machines a DC (Direct Current) machine by another British scientist William Sturgeon in the year 1832. But this machine was very expensive and wasn't used for any practical purpose due to the absence of energy sources.

The first induction motor was invented by the eminent Nikola Tesla in 1887. The invention of cage-rotor induction motor in 1889 and the three-limb transformer in 1890 by Mikhail Dolivo-Dobrovolsky have contributed a lot towards industrial developments. Tesla and Westinghouse have developed two phase, 60 Hz, wound rotor induction motors in their early stages and B. G. Lamme in 1893 come up with first practical induction motor with a rotating bar winding rotor. The General Electric Company (GE) began developing three-phase induction motors in 1891. By 1896, General Electric and Westinghouse signed a cross-licensing agreement for the bar-winding-rotor design, later called the squirrel-cage rotor. Induction motors find their significant place in industrial drive applications and out rule the dc motors since ac power is used in generation, transmission and distribution.

The advancements in the technology have resulted in inventions and innovations such that the mounting dimensions of a 7.5 HP motor of 1897 has the same dimension as that of a 100 HP induction motor. The range of power available in market are from 0.25 – 500 HP. The general makes are Siemens, Crompton grieves, Kirloskar, Bharat Bijlee, Hindustan, Marathon, ABB etc

1.2 CONSTRUCTION OF THREE PHASE INDUCTION MOTOR:

An Induction motor mainly consists of two parts: (a) Stator and (b) Rotor

(a) Stator:

(i) Stator core is made of laminated silicon steel stampings and has slots and teeth on its inner periphery to house stator windings as shown in Fig.1.1 a, b and c. The stampings are 04.to 0.5 mm thick.
(ii) Stator carries a 3-phase winding displaced in space by 120° electrical
(iii) The 3-phase winding is either star or delta connected and is fed from 3-phase supply
(iv) The radial ventilating ducts are provided along the length of the stator core
(v) For higher capacity machines the stator core is segmented as shown in Fig 1.1 c.

(a) (b) (c)

Fig.1.1: a) Stator core stampings for a low rating machine, b) yoke and Stamping of three phase induction motor, c) Stator core stampings for a higher rating machines

(b) Rotor:

(i) Rotor comprises a cylindrical laminated iron core, with slots on outer periphery
(ii) Like stator, rotor laminations are punched in one piece for small Machine as shown in Fig 1.2.
(iii) In larger machine the laminations are segmented.
(iv) If there are ventilating ducts on the stator core, an equal number of such ducts is provided on rotor core.

(a) (b)

Fig.1.2: a) Rotor stampings, b) Structure of the laminated rotor

There are two types of rotor constructions:

a) Squirrel cage rotor
b) Slip ring or wound rotor

a) Squirrel Cage Rotor:

i. This rotor consists of a cylindrical laminated core with parallel slots

ii. Rotor slots are usually not quite parallel to the shaft but for reducing the magnetic hum and locking tendency rotor slots have slight skew

iii. In rotor slots heavy copper, aluminium or alloy bars are housed

iv. Rotor bars are permanently short circuited at the ends. This limits the insertion of any external resistance.

Now let us have a look into the exploded view of the induction motor shown in Fig.1.3. The rotor is die casted in aluminium and it is short circuited at both the ends. The picture reveals that this aluminium ring shorts all the coils and that is why it looks like a squirrel cage. Then we could see some coils or windings inside called stator windings which form various poles corresponding to A phase, B phase and C phase which are wound on the circumference of this stationary system. This is called the yoke where the stator is anchored on to the ground through these foot plates. i.e why we say that once the poles are fixed in space they do not rotate and what makes the flux to rotate is basically by virtue of the 3-phase source that you supply which are 120° apart. The rotor is inserted into this stator and then the end plates are fitted in through these holes and bearing that will cause a full smooth rotation of the shaft. Beyond this face plate there is fan mounted on the shaft and there is another enclosure for covering the fan. This fan is to provide cooling for the machine i.e. an air-cooled induction machine. All these windings are brought out as three terminal points A phase, B phase, C phase and all neutrals connected together for one or other end of the coil of the various phases. Of course for various ratings the sizes will be different.

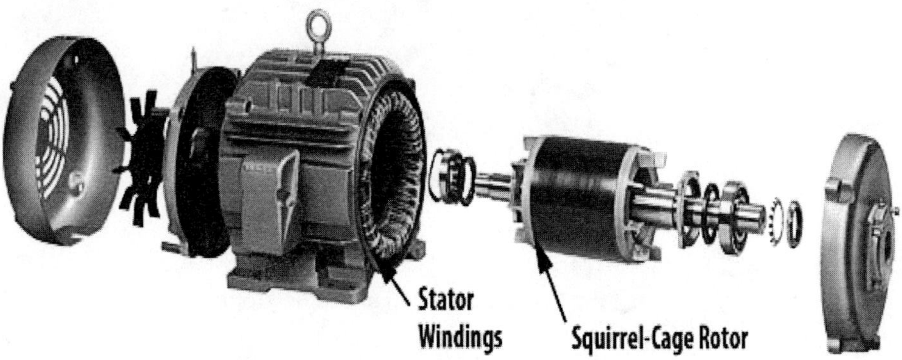

Fig.1.3: Exploded view of Squirrel cage induction motor

Advantages of squirrel cage induction rotor are:

- ✓ Its construction is very simple and rugged.
- ✓ As there are no brushes and slip ring, these motors require less maintenance.

Applications:

Squirrel cage induction motor is used in lathes, drilling machine, fan, blower printing machines etc

b) Slip Ring or Wound Rotor:

(i) The rotor is wound for the same number of poles and number of phases as that of stator
(ii) Rotor winding is either star or delta but star connection is preferred
(iii) The three star terminals are connected to three brass slip ring mounted on rotor shaft as shown in Fig.14.
(iv) These slip rings are insulated from rotor shaft
(v) Slip rings connected with brushes and three brushes can further be connected externally to 3- variable rheostats.
(vi) This makes possible to introduce additional resistance in the rotor circuit during starting period

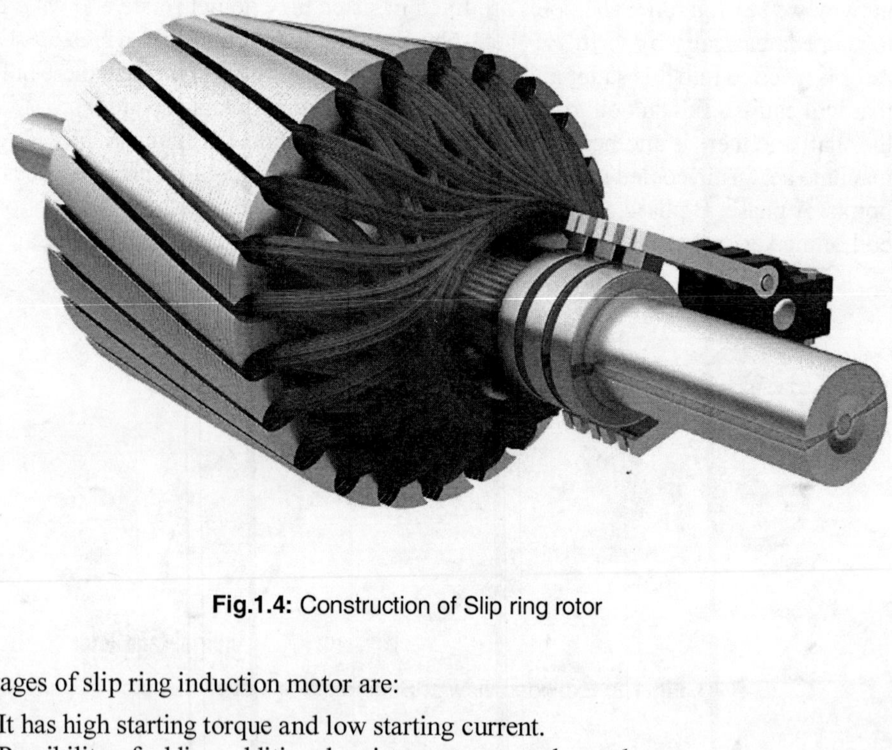

Fig.1.4: Construction of Slip ring rotor

Advantages of slip ring induction motor are:

- ✓ It has high starting torque and low starting current.
- ✓ Possibility of adding additional resistance to control speed.

Applications:

Slip ring induction motor are used where high starting torque is required i.e. in hoists, cranes, elevator etc.

Table 1: Difference between Slip Ring and Squirrel Cage Induction Motor

S. No	Feature	Slip ring or phase wound Induction motor	Squirrel cage induction motor
1.	Construction	Complicated due to presence of slip ring and brushes	Very simple
2.	Windings	Similar windings in stator and rotor.	Short circuited rotor bars
3.	Rotor resistance	Resistance can be added by using slip ring and brushes	it's not possible to add external resistance
4.	Starting torque	High starting torque can be obtained by adding external resistance	Staring torque is low and cannot be improved
5.	Brushes and slip rings	Slip ring and brushes are present	Slip ring and brushes are absent
6.	Maintenance	Frequent maintenance is needed	Less maintenance only required
7.	Cost	More	Cheap
8.	Usage	Rarely used in industry	Due to its simple construction and low cost. Hence, widely used
9.	Efficiency	Rotor copper losses are high and hence less efficiency	Less rotor copper losses and hence high efficiency
10.	Speed control	Speed control by rotor resistance method is possible	Speed control by rotor resistance method is not possible
11.	Applications	Slip ring induction motor are used where high starting torque is required i.e. in hoists, cranes, elevator etc.	Squirrel cage induction motor is used in lathes, drilling machine, fan, blower printing machines etc.

1.3 WORKING PRINCIPLE

Induction motors are very popular since 80% of the prime movers all over the world use induction motors for various applications. Let us began our discussion by an initial comparison of the induction motor with the 3-phase transformers. In the case of 3-phase transformers, the energy moves from the electrical side into the magnetic domain and then back again into the electrical domain in the secondary. A similar energy transfer happens from the primary side coil called the stator to the secondary side coil called the rotor, another domain called the mechanical domain, in induction motor. We can tap the energy in the electrical domain or in the mechanical domain or both depending upon the type of the motor.

Electrical power is converted to mechanical power in the rotating part of an electrical motor with the help of commutator hence a dc motor can be called as conduction motor. However, in ac motor the rotor does not receive electric power by conduction but by induction in exactly the same way as the secondary of a 2-winding transformer receives its power from the primary i.e.

why such motors are known as induction motors. In fact, an induction motor can be treated as a rotating transformer i.e. one in which primary winding is stationary with the secondary winding free to rotate. Thus, it works on the principle of induction where electro-magnetic force (emf) is induced in to the rotor conductors when rotating magnetic field of stator cuts the stationary rotor conductors. As discussed in the last section, the rotor is short circuited in squirrel cage motor and cannot take any electrical energy from the rotor. So, all the electrical energy can be taken only from the mechanical domain. i.e the energy path is electrical domain – magnetic - mechanical domain and in the case of the other category the wound rotor induction motor or the slip ring induction motor the electrical domain, energy enters the machine goes into the mechanical domain and the energy can be tapped out either in the electrical domain of the rotor or the mechanical domain of the rotor.

According to Faraday's laws of electromagnetic induction, a voltage ($e = B\ell v$) is induced on the rotor conductors and thereby a current is induced in the short-circuited rotor conductors by virtue of the relative motion of the rotor conductors with respect to the field inside the machine. By Lorentz law ($F = BIL$) there is a force which is induced on the rotor conductors and the accumulated force on all the rotor conductors create a torque and makes the rotor move.

1.4 PRODUCTION OF ROTATING MAGNETIC FIELD

When a 3-Ø stator winding displaced in space by 120° electrical is energized by a 3- Ø supply having 120° displaced in space, a rotating magnetic field sets up in the stator. But the important concept now is that how do we bring about a relative motion between the field in the motor and the rotor conductors. Now, to achieve that we use a 3-phase source with same effective amplitudes displaced in space by 120°. These currents flow through coils in the stator of the motor wherein the coils are physically located in space by 120°, a combination of three phase supply given to three phase winding produces a rotating magnetic field. So let us now see how this rotating magnetic field is created in the stator.

Let f be the frequency of the stator supply. i.e. f cycles/s.

Let P be the number of poles/phase of the motor.

Time taken by the rotor to make full 360° or 1 revolution = 1/f s.

Hence, the number of revolutions made by the rotor in a minute or 60s for one pole pair = 60f

The number of revolutions per minute = 120f/P

This speed is called as the synchronous speed of the machine which is dependent on the number of poles and frequency of the supply. The number of poles is always an even number. The synchronous frequency increases with supply frequency and decreases with increase in the number of poles.

Let us try to understand the theory behind the production of rotating magnetic field. For that, imagine the stator of an electric motor where the three-phase winding is physically distributed in the stator core displaced in space by 120°. Let us assume the phasors of each phase as shown in Fig.1.5.

Although the vector sum of three currents in a balanced three-phase system is zero at any instant, the resultant of the magnetic fields produced by the currents is not zero rather it will have a constant non-zero value rotating in space with respect to time. The magnetic flux produced by the current in each phase can be represented by:

$$\Phi_R = \Phi_m \sin(\omega t)$$
$$\Phi_Y = \Phi_m \sin(\omega t - 120°)$$
$$\Phi_B = \Phi_m \sin(\omega t - 240°)$$

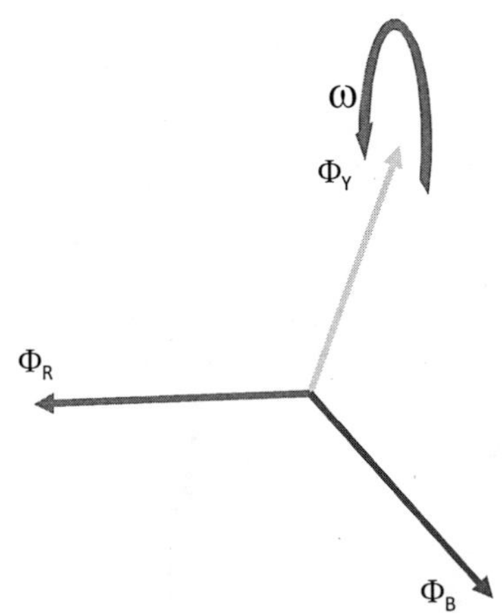

Fig.1.5: Phasor representation of stator fluxes

Where, Φ_R, Φ_Y and Φ_B are the instantaneous flux of corresponding Red, Yellow and Blue phase winding, Φ_m amplitude of the flux wave. The variation of flux w.r.t space can be represented as shown in Fig.1.6.

Let us consider the first instant when θ = 0° as shown in Fig.1.6.

The instantaneous values of R, Y and B phase fluxes are

$$\Phi_R = \Phi_m \sin(0) = 0$$

$$\Phi_Y = \Phi_m \sin(-120°) = (-0.5\sqrt{3})\Phi_m$$
$$\Phi_B = \Phi_m \sin(-240°) = (0.5\sqrt{3})\Phi_m$$

The resultant of these fluxes at 0° i.e. Φ_r is $1.5\Phi_m$ which is shown in Fig.1.7.

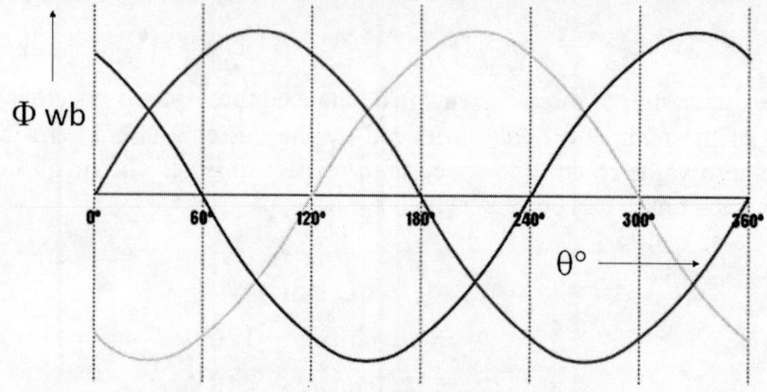

Fig.1.6: Variation of stator flux w.r.t space

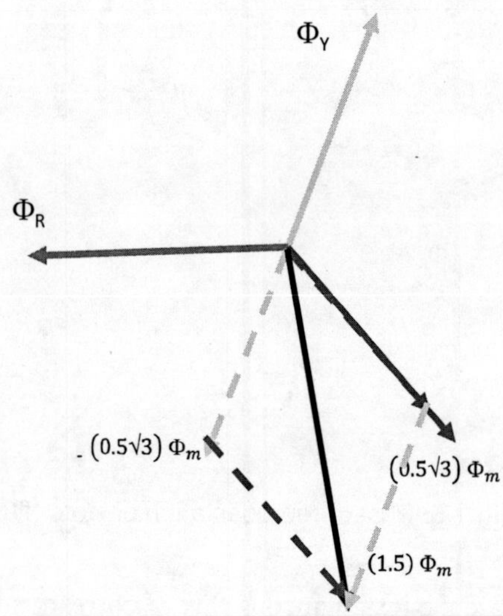

Fig.1.7: The resultant flux at 0°

Now, let us consider the 2nd instant where θ = 60° as shown in Fig.1.6.

The instantaneous values of R, Y and B phase fluxes are

$$\Phi_R = \Phi_m \sin(60) = 0.5\sqrt{3}\Phi_m$$

$$\Phi_Y = \Phi_m \sin(60 - 120^\circ) = (-0.5\sqrt{3})\Phi_m$$

$$\Phi_B = \Phi_m \sin(60 - 240^\circ) = 0$$

The resultant of these fluxes at instant 2 is $1.5\Phi_m$ which is shown in Fig.1.8. Also, this phasor is rotated in the clockwise direction through 60°.

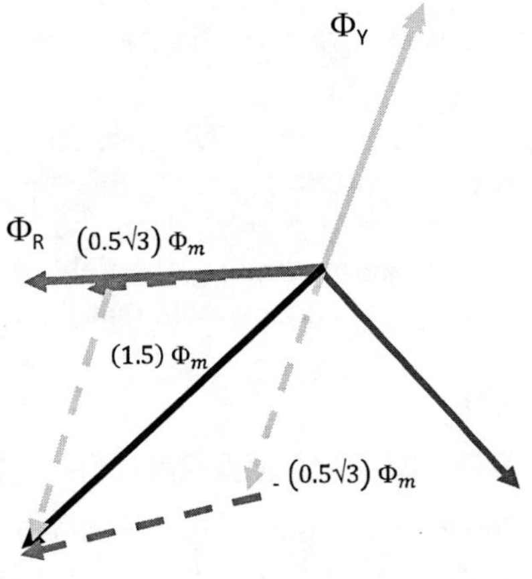

Fig.1.8: The resultant flux at 2nd instant

Let us consider the 3rd instant where ωt = 120° as shown in Fig.1.6.

The instantaneous values of R, Y and B phase fluxes are

$$\Phi_R = \Phi_m \sin(120) = 0.5\sqrt{3}\Phi_m$$

$$\Phi_Y = \Phi_m \sin(120 - 120°) = 0$$

$$\Phi_B = \Phi_m \sin(120 - 240°) = -0.5\sqrt{3}\Phi_m$$

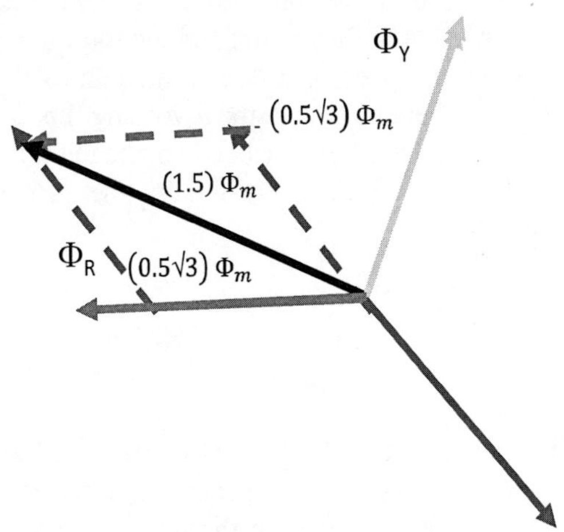

Fig.1.9: The resultant flux at 3rd instant

The resultant of these fluxes at instant 3 is 1.5 Φ_m which is shown in Fig.1.9. Here the resultant flux is rotated further 60°. In this way we can prove that due to balanced supply applied to the three - phase stator winding a magnetic field rotating or revolving at constant speed is established.

In general, for P poles, the rotating field makes one revolution in P/2 cycles of current. Referring to the waveform of three phases, the time instant 4 represents the completion of one quarter cycle of alternating current from the time instant 1. During this one-quarter cycle, the field has rotated through 90°. At one complete cycle of current from the origin, the field has completed one revolution. Therefore, for a 2-pole stator winding, the field makes one revolution in one cycle of current. In a 4-pole stator winding, it can be shown that the rotating field makes one revolution in two cycles of current.

1.5 DIRECTION OF ROTATING MAGNETIC FIELD

The phase sequence of the three-phase voltage applied to the stator winding is R-Y-B as shown in Fig.1.6. If this sequence is changed to R-B-Y, it is observed that the direction of rotation of the field is reversed i.e., the field rotates counter clockwise rather than clockwise. However, the number of poles and the speed at which the magnetic field rotates remain unchanged. Thus it is necessary to change the phase sequence in order to change the direction of rotation of the magnetic field. For a three-phase supply, this can be done by interchanging any two of the three lines. Therefore, the direction of rotation of a 3-phase induction motor can be reversed by interchanging any two of the three motor supply lines.

1.6 ROTOR SPEED AND SLIP

Now the synchronously revolving flux sweep past the air gap and cuts the short-circuited rotor conductors. This induces a current in the rotor circuit. The direction of this current is such as to oppose the relative speed between the stator flux and the rotor speed. Hence the rotor rotates at a speed less than the synchronous speed in a direction same as that of stator flux reducing the relative speed. The rotor current interacts with the revolving flux in the stator producing a torque. The slip in an induction motor is the difference between the main flux speed and their rotor speed. The symbol *s* represents the slip. It is expressed by the percentage of synchronous speed. Mathematically, it is written as,

$$\% s = \frac{N_s - N_r}{N_s} \times 100$$

The value of slip at full load varies from 6% in the case of a small motor and 2% in a large motor. The induction motor never runs at synchronous speed. The speed of the rotor is always less than that of the synchronous speed. If the speed of the rotor is equal to the synchronous speed, no relative motion occurs between the stationary rotor conductors and the main field. Then no EMF induces in the rotor and zero current generates on the rotor conductors. The electromagnetic torque is also not induced. Thus, the speed of the rotor is always kept slightly less than the synchronous speed. The speed at which the induction motor work is known as the slip speed. The difference between the synchronous speed and the actual speed of the rotor is known as the slip speed. In other words, the slip speed shows the relative speed of the rotor concerning the speed of the field. The speed of the rotor is slightly less than the synchronous speed. Thus, the slip speed expresses the speed of the rotor relative to the field.

Therefore, the rotor speed is obtained as:

$$N_r = N_s(1 - s)$$

Slip plays an essential role in the induction motor. As we know, the slip speed is the difference between the synchronous and rotor speed of the induction motor. The emf induced in the rotor because of the relative motion, or we can say the slip speed of the motor. So,

$$e_r \propto N_s - N_r$$

The rotor current is directly proportional to the rotor induced emf

$$i_r \propto e_r$$

Torque is proportional to the rotor current. Therefore,

$$T = K\ (N_s - N_r) \propto \mathrm{s}$$

i.e. the torque is directly proportional to slip.

The above equation shows that the torque induced on the rotor is directly proportional to the slip of the induction motor. The high value of slip induces a greater emf in the rotor. This EMF develops a heavy torque on the rotor conductors.

The value of the slip is adjusted by considering the load on the motor. For full-load, a high value of torque is required. This can be achieved by increasing the amount of the slip and reducing the speed of the rotor. The slip of the motor is kept low when the induction motor is running at no-load. The small slip produces a small torque on the motor. The value of the induction motor slip is adjusted according to the requirement of the driving torque at the normal working condition.

1.7 SIGNIFICANCE OF SLIP

Let us take a 3 phase, 4 pole, 50 Hz induction motor. Now for this induction motor calculate the frequency of the rotor currents f_r for the following conditions: 1) at standstill meaning the rotor shaft is not moving; 2) motor is running at 500 rpm in the same direction as the field; 3) motor is running at 500 rpm however in opposite direction as the field rotates and 4) what is the condition when the motor is running at 2000 rpm in same direction as field.

Let us understand the operation of the motor under these four running conditions.

Synchronous speed = 120 f/P = 120×50/4 = 1500 rpm.

case 1: at standstill, speed of rotor = 0 rpm.

Slip = 1500-0/1500 = 1

Rotor current frequency = sf = 50 Hz.

Case 2: motor turns at 500 rpm in the same direction as the field

slip = (1500 – 500) / 1500 = 1000 / 1500 = 2 / 3 = 0.66.

rotor frequency = sf = 33.33 Hz.

When the rotor speed is 500 rpm in the same direction as the field the mode of operation is motoring in its regular manner.

Case 3: The motor turns at 500 rpm in opposite direction with respect to the field.

Let us say counter clockwise rotation as positive, clockwise as negative, so if the stator flux is rotating in a counter clockwise or anticlockwise direction (positive) and say that the motor is rotating in clockwise direction. Then the slip is (1500 – (- 500)) / 1500 =2000/1500 = 1.33. The slip is greater than 1. So, if the slip is greater than 1 it is breaking operation and whenever the slip is between 0 and 1 this is normal motoring operation.

Case 4: The motor is rotating at 2000 rpm in the anticlockwise direction itself along the stator field. This means that the mechanical shaft speed is greater than the synchronous speed.

Slip = 1500-2000/1500 = - 0.33.

Now the rotor frequency = - 0.33×50 = -16.66 Hz.

Consider a vehicle in which the induction motor is used as the drive. Let us assume this vehicle is moving uphill. The rotating m.m.f is moving at synchronous speed and the shaft is trying to catch up with it and it is motoring and it is pulling the vehicle. And once the vehicle has gone uphill and then it starts going downhill the motor is also going to be aided by the gravity, so the motor is also going to rotate not only by the electrical energy that is being supplied but also is going to get aided by the gravity and if the gravitational acceleration is high then the shaft speed can go beyond the synchronous speed. In this case the inertial energy which is there in the mass of the vehicle is put back to the supply i.e. the motor is acting like a generator, and then it is called an induction generator.

1.8 ROTOR CURRENT FREQUENCY

The frequency of the applied the voltage and current produced in the stator are the same and is given by,

$$f = \frac{N_s \times P}{120}$$

The rotor induced emf is proportional to the relative speed between stator flux and rotor speed. i.e., on the slip. Thus, the rotor frequency is given by,

$$f_r = \frac{(N_s - N_r) \times P}{120}$$

$$\frac{f_r}{f} = \frac{\frac{(N_s - N_r) \times P}{120}}{\frac{N_s \times P}{120}} = \text{s}$$

$$f_r = s\,f$$

When the rotor is stationary, i.e., $N_r = 0$, then, s =1. Hence, the frequency *(f_r)* of the rotor current is the same as that of the supply frequency *(f)*.

$$f_r = f$$

When the rotor picks up speed, the relative speed between the rotating magnetic field and the rotor decreases. As a result of this, the slip (s) and hence the rotor current frequency decreases.

At synchronous speed, i.e., $N_r = N_S$, s = 0 and hence the frequency of the rotor current is zero.

$$f_r = 0$$

1.9 ROTOR EMF

By Faraday's laws of electromagnetic induction, induced emf

$$e = N\frac{d\emptyset}{dt} = N\frac{d}{dt}(\emptyset_m \sin \omega t) = N\,\emptyset_m \omega \cos\omega t$$

i.e. magnitude of induced emf in the stator (e_s) is proportional to $N_s\ \emptyset_m f$

The induced emf in the rotor (e_r) is proportional to $N_r\ \emptyset_m s f$

$$\frac{e_s}{e_r} = \frac{K_{ws} N_s}{K_{wr}\ s\ N_r}$$

Where K_{ws} and K_{wr} are winding factor for stator and rotor respectively.

UNIT SUMMARY

The aim of this chapter is to understand the construction and working principle of the induction motor. The induction machine has such a robust and straight forward construction. The principle of operation of the motor conveys clearly why induction motor is called asynchronous machine. Also, discussed about the revolving field and how it helps in starting the three-phase induction motor. The significance of positive, negative and zero slip are discussed in detail.

EXERCISES

1. A 3-phase 8 pole 50 Hz induction motor runs at 740 r.p.m. find its % slip.

2. A 12 pole 3-phase alternator driven at speed of 500 r.p.m. supplies power to an 8 pole 3 φ induction motor. If the slip of motor is 0.03 p.u, calculate the speed.

3. A 3-phase 4 pole induction motor is supplied from 3φ 50Hz ac supply. Find synchronous speed rotor speed when slip is 4% the rotor frequency when runs at 600r.p.m.

4. A 12 pole 3-phase alternator is coupled to an engine running at 500r.p.m. If supplied a 3φ induction motor having full speed of 1440r.p.m. Find the %age slip, frequency of rotor current and no of poles of rotor.

5. The rotor of 3-phase induction motor rotates at 900 r.p.m. when stator is connected to 3φ supply, find the rotor frequency.

6. A 3-phase 50Hz induction motor has a full load speed of 960 r.p.m find slip, No. of poles, Frequency of rotor induced e.m.f, Speed of rotor field w.r.t. rotor structure, Speed of rotor field w.r.t. Stator structure, Speed of rotor field w.r.t. stator field, Speed of rotor field w.r.t stator field is zero.

7. A 3-phase, 400V wound rotor has delta connected stator winding and star connected rotor winding. The stator has 48 turns/phase while rotor has 24 turns per phase. Find the stand still or open circuited voltage across the slip rings.

Multiple Choice Questions

1. "The direction of induced current in the rotor is so as to oppose the cause producing it". This is given by

   a. Faraday's law

   b. Lenz's law

   c. Electromagnetic law

   d. Ampere's law

2. If the rotor of the machine catches the speed of the rotating magnetic field, then the motor will

   a. Rotate at the speed of rotating magnetic field

   b. Rotate at double the speed of the rotating magnetic field

   c. Eventually stop

d. None of these

3. Slip speed of the motor decides the magnitude of the induced emf and the rotor current, which in turn decides the torque produced. If Ns is the synchronous speed and N is the motor speed in rpm, then the sip speed is given by

   a. Ns

   b. Ns – N

   c. Ns + N

   d. N-Ns

4. At star, the slip of the induction motor is

   a. 1

   b. 0

   c. 0.5

   d. None of these

5. Slip of an induction motor will be zero,

   a. At start

   b. When machine is rotating at its maximum speed

   c. When it is rotating at half of the maximum speed

   d. Cannot be zero

6. A 4 pole, 3-phase, 50 Hz star connected induction motor has a full load slip of 5%. Then full load speed of the motor is

   a. 1500

   b. 1400

   c. 1405

   d. 1425

7. A 50 Hz, 3-phase slip ring induction motor, has 6 poles on stator and 4 poles on rotor. Then the machine will run at

   a. 1000 rpm

   b. 1500 rpm

c. 1400 rpm

d. Machine will not run at all

8. A 3-phase, 50 Hz squirrel cage induction motor, has 6 poles on stator and 4 poles on rotor. Then the machine will run at

a. 1000 rpm

b. 1500 rpm

c. 1750 rpm

d. Machine will not run

9. The rotor frequency of a 3-phase induction motor is

a. Slip times the supply frequency

b. 1 / slip times the supply frequency

c. Equal to the supply frequency

d. None of these

10. Nature of the rotor power factor in running condition is always

a. Leading

b. Lagging

c. Either leading or lagging

d. Neither leading nor lagging

Answers of Multiple Choice Questions

1. b. Lenz's law
2. c. Eventually stop
3. b. Ns – N
4. a. 1
5. d. Cannot be zero
6. d. 1425
7. d. Machine will not run at all
8. a. 1000 rpm
9. a. Slip times the supply frequency
10. b. Lagging

Short and Long Answer Type Questions

Long Answer Questions with Answers

1. Explain the working principle of three phase induction motor.

- When a 3-Ø stator winding displaced in space by 120° electrical is energized from a 3- Ø supply having 120° time displacement a rotating magnetic field is setup in the stator.
- This rotating magnetic field revolving at synchronous speed is established in the air gap between the stator and rotor. $N_s = 120. f/P$ Where N_s represents the synchronous speed, f represents the supply frequency and P represents the number of poles in the machine.
- This rotating magnetic field passes through the air gap and cuts the stationary rotor conductors.
- Due to the relative speed between the rotating flux and the stationary rotor EMFs are induced in the rotor conductors.
- If the rotor conductors are short circuited, currents start flowing in the rotor conductor.
- According to Lenz's law the direction of the induced current is such that it opposes the cause.
- Cause is the relative speed between the rotating field and stationary rotor.
- Hence, the rotor has a tendency to reduce its relative speed.
- So, rotor begins to move in the direction of rotating field and continues towards achieving synchronous speed and the machine runs at a speed near but below synchronous speed (the rotor slips back) depending upon load on shaft.
- As the speed of rotor reaches synchronous speed (speed of the stator magnetic field) relative speed becomes zero. Hence, no emf, no current and therefore no torque at synchronous speed. Hence rotor never reaches to synchronous speed.

2. What happens when the rotor of the induction motor runs at synchronous speed?

If the rotor attains synchronous speed, the relative speed between stator flux and the rotor conductors become zero resulting no induced emf and induced current in the rotor conductors. Hence, rotor flux fails to be established hindering induced torque. This slow down the rotor due to friction as shown in the Fig.

Zero friction and other losses is a hypothetical / imaginary condition. there cannot be any motion without friction. In case of motor, there are other losses like losses due to air resistance, iron losses, copper losses etc. Motor approaching synchronous speed is possible only if it is having zero friction, zero air resistance, zero iron loss (eddy current loss and hysteresis loss in the core), copper loss (I square R loss in the rotor) etc. An induction motor can thus speed up to near synchronous speed but it can never reach synchronous speed.

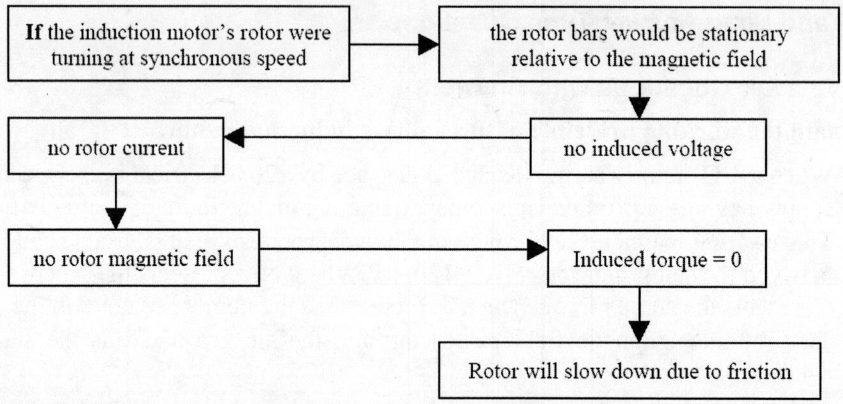

Short answer Questions:

1. Why are 3-phase induction motors very popular as drives for industrial applications?
2. What are the various types of 3-phase induction motors as per the rotor construction?
3. List the differences between squirrel cage and slip ring rotor.
4. Define slip of induction motor.
5. A 3-phase induction motor does not run at synchronous speed. Why?
6. How to change the direction of rotation of induction motor?
7. Write the application of slipring induction motor.

Long answer Questions:

1. With the help of diagrams, explain how a rotating magnetic field is produced in the air gap of a 3phase induction motor.
2. Explaintheprincipleofoperationof3-phaseinductionmotor.
3. Explainwithneatdiagramofconstructionofsquirrelcageinductionmotor.

Numerical Problems

1. A 3-phase, 50 Hz induction motor has 8 poles and operates with a slip of 4 % at a certain load. Determine the frequency of the rotor current. [2Hz]
2. A 3 φ 4 pole 50 Hz induction motor runs at 1460 r.p.m. find its % slip. [2.667%]
3. A 12 pole 3 φ alternator driven at speed of 500 r.p.m. supplies power to an 8 pole 3 φ induction motor. If the slip of motor is 0.03 p.u, calculate the speed. [727.5 rpm]
4. A 3-φ 4 pole induction motor is supplied from 3φ 50Hz ac supply. Find
   a. synchronous speed
   b. rotor speed when slip is 4%
   c. the rotor frequency when runs at 600r.p.m. [1500 rpm; 1440 rpm; 30Hz]
5. A 3 φ 50Hz induction motor has a full load speed of 960 r.p.m
   a. find slip
   b. No of poles
   c. Frequency of rotor induced e.m.f

d. Speed of rotor field w.r.t. rotor structure
e. Speed of rotor field w.r.t. Stator structure
f. Speed of rotor field w.r.t. stator field [4%; 6; 2Hz; 40 rpm; 1000 rpm; 0 rpm]

PRACTICAL

1. Identify various parts of the 3-phase induction motor and list various materials used for constructing.
2. Identify various applications of induction motor and note the specifications of the motor in various applications. Note the following: voltage, current, torque, speed etc.
3. Observe the three-phase supply voltage through a sensor
4. Measure the voltages applied in each phase of the three-phase induction motor.
5. How to change the direction of rotation of the rotor?
6. Calculate the Slip speed of the given induction motor experimentally.

KNOW MORE

1. Working of Induction motor
   https://www.youtube.com/watch?v=AQqyGNOP_3o&ab_channel=Lesics
2. How to manufacture a three-phase induction motor
   https://www.youtube.com/watch?v=esYVPt77uO4&ab_channel=NideGroup
3. Mass production of Induction motor
   https://www.youtube.com/watch?v=qsW-sCsc&ab_channel=SMTWindingEquipment
4. Disassembling three phase induction motor
   https://www.youtube.com/watch?v=hd9e7cx_qkc&ab_channel=SkillFootage
5. Revolving magnetic field
   https://www.youtube.com/watch?v=vMu6DmfKHTs&ab_channel=learnchannel

REFERENCES AND SUGGESTED READINGS

1. P.S. Bimbhra, Electric Machines, Khanna Book Publishing Co., New Delhi (ISBN: 978- 93-6173- 294)
2. Mittle, V.N. and Mittle, Arvind., Basic Electrical Engineering, McGraw Hill Education New Delhi, ISBN :9780070593572
3. Kothari, D. P. and Nagrath, I. J., Electrical Machines, McGraw Hill Education. New Delhi, ISBN:9780070699670
4. Bhattacharya, S. K., Electrical Machines, McGraw Hill Education, New Delhi, ISBN:9789332902855
5. Theraja, B.L., Electrical Technology Vol-II (AC and DC machines), S.Chand and Co. Ltd., New Delhi, ISBN : 9788121924375
6. Sen, S. K., Special Purpose Electrical Machines, Khanna Publishers, New Delhi, ISBN: 9788174091529

7. Janardanan E. G, Special Electrical Machines, Prentice Hall India, New Delhi ISBN: 9788120348806
8. Hughes E., Electrical Technology, ELBS
9. Cotton H., Electrical Technology, ELBS

Dynamic QR Code for Further Reading

2 THREE-PHASE INDUCTION MOTOR - T vs N Characteristics

UNIT SPECIFICS

Through this module we have discussed the following aspects:

- *Consideration as a rotating transformer representing equivalent circuit;*
- *Variation of torque w.r.t slip (speed);*
- *Phasor diagram;*
- *Power factor at starting and running conditions;*
- *Power flow diagram;*
- *Four quadrant operation;*
- *Problems.*

For the purpose of fostering more curiosity and creativity as well as enhancing problem-solving abilities, the themes' practical applications are covered.

Along with a large number of multiple-choice questions and questions with short and long answers divided into two categories based on Bloom's taxonomy's lower and higher orders, the unit also includes assignments through a number of numerical problems, a list of references, and suggested readings that one can use as practice materials. It is significant to notice that several portions of the website feature QR codes that can be scanned for further information on a variety of interesting topics.

Following the relevant practical exercise, a "Know More" section is provided, which expands upon the information discussed. The present section has been meticulously crafted to ensure that the additional material included herein becomes advantageous to the readers of the book. This section primarily emphasises the introductory aspects, notable facts, analogies, historical background, key observations and findings, chronological progression from the inception of the subject to the present, practical applications in everyday life and industry, relevant case studies pertaining to environmental, sustainability, social, and ethical concerns, and lastly, topics that provoke curiosity and inquiry within the unit.

RATIONALE

Understanding the relationship between torque and slip (speed) is crucial for analysing the performance of induction motors. The torque-slip characteristics provide insights into the motor's ability to generate torque at different operating conditions. Studying these characteristics helps in motor selection, load matching, and predicting the motor's behaviour under varying load conditions.

Induction motors exhibit different torque characteristics during different operating conditions. Studying the starting torque is important to ensure that the motor can overcome the inertia and start the load. Full load torque indicates the maximum continuous torque that the motor can deliver under normal operating conditions. Understanding the relationship between starting torque, full load torque, and maximum torque helps in motor sizing and ensuring reliable operation.

The induction motor can be conceptualised as a generalised transformer, wherein electrical power is converted to mechanical power. The understanding of the phasor diagram depiction facilitates the analysis of the electrical and magnetic fields within the motor, along with the examination of power flow and the relationships between voltage, current, and power factor.

Induction motors can operate in all four quadrants of the torque-speed plane, allowing for both motoring and generating modes. Understanding the four-quadrant operation is essential for applications such as regenerative braking or motor-generator sets. The power flow diagram illustrates the transfer of power between the electrical and mechanical domains during different motor operating conditions.

PRE-REQUISITES

Machines I: Sem III

Physics: Electromagnetism (Class X)

UNIT OUTCOMES

List of outcomes of this unit is as follows:

U2-O1: Rotating transformer analogy to deduce equivalent circuit

U2-O2: Derive the torque equation and describe the T-N characteristic

U2-O3: Deduce phasor diagram and power factor

U2-O4: Describe power flow in the motor

U2-O5: Explain four quadrant operation of the motor as a drive

Unit-2 Outcomes	EXPECTEDMAPPINGWITHCOURSEOUTCOMES (1-WeakCorrelation;2-Mediumcorrelation;3-StrongCorrelation)				
	CO-1	CO-2	CO-3	CO-4	CO-5
U2-O1	3	-	-	-	-
U2-O2	3	-	-	-	-
U2-O3	-	3	-	-	-
U2-O4	-	3	-	-	-
U2-O5	-	3	-	-	-

2.1 INTRODUCTION

In the realm of electrical motors, induction motors play a significant role due to their efficiency, reliability, and versatility. Understanding the characteristics of torque versus slip (speed) is crucial to comprehend the behaviour of these motors across various operational scenarios. Moreover, exploring the different types of torques - starting, full load, and maximum - and their relationships provides valuable insights into the performance capabilities of induction motors. To deepen our understanding, we will also examine the induction motor as a generalized transformer, incorporating phasor diagrams to visualize its operation.

Additionally, we will delve into the concept of four-quadrant operation and power flow diagrams to illustrate the power exchange between the motor and its load. Through this comprehensive exploration, we will gain a comprehensive understanding of the characteristics, operation, and control of induction motors, contributing to our knowledge of electric motor systems.

2.2 RELATIONSHIP BETWEEN MECHANICAL DEGREES AND ELECTRICAL DEGREES

As discussed in unit I the stator of the induction motor may be wound for 2 poles, 4 poles, 6 poles etc. Consider a 2 pole 3 phase winding, for such a winding, the mechanical rotational frequency of the magnetic field is f_0 Hz (say 50 times/s).

A 4 pole 3 phase winding exhibits a rotational frequency of $f_0/2$ Hz, i.e., 25 Hz. The electrical phenomenon for this case with a mechanical rotation is shown in Fig.2.1. A 6 pole 3 phase winding exhibits a rotational frequency of $f_0/3$ Hz and so on. So, the maximum mechanical frequency of the rotor is for 2 poles. As the number of poles increases the speed of the machine become slower. This reduction in the mechanical speed is due to the fact that whatever electrical and electromagnetic phenomenon witnessed for a mechanical angle range 0-2π for a 2-pole machine is fully seen in the electrical angle range 0-2π/p for a 2p pole winding. Here p represents the number of pole pairs.

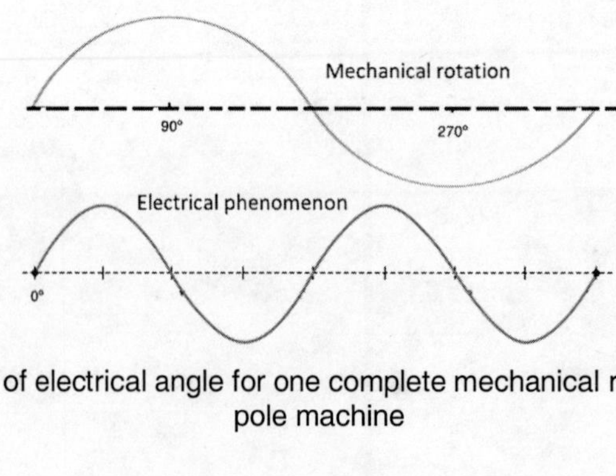

Fig.2.1: Representation of electrical angle for one complete mechanical rotation of the rotor for a 4-pole machine

i.e., one cycle of the electrical phenomenon like voltage, current and flux density for a 2-pole winding gets completed within 2π/p radians. Therefore, mechanical angle of 2π radians is equivalent to 2π/p electrical radians.

The equivalent circuit for the electrical phenomena is created at the mains frequency. The mechanical quantities like shaft torque, speed etc are computed at the mechanical speed.

Example:

A 50 Hz induction motor runs at 1450 rpm. The power output is 3600 W. Determine (a) number of poles, (b) slip, (c) torque developed.

Nearest synchronous speed of this motor is 1500 rpm.

$$Ns = 120.\frac{f}{P} = 120.\frac{50}{P}$$

P = 4

Slip = $N_s - N_r / N_s$ = 1500 – 1450 / 1500 = 0.033 = 3.33%

Mechanical power = 3600 W = τ. ω W

Torque τ = 3600/ω = 3600/(2π N_r/60) = 23.7 Nm.

a) EMF Equation

Consider any one phase of the induction motor with 2 poles. Let the number of turns/phase be T. There is a rotating magnetic field in the air gap which when cut by the stator conductors give rise to an alternating voltage. This induced emf will counter balance the applied voltage under ideal conditions. The flux distribution in the air gap is sinusoidal.

For a conductor of length L rotating with an angular velocity ω making an angle θ as shown in Fig.2.2, the induced voltage = B sin θ L (rω)

For a coil (2 conductors), the induced voltage = 2 B sin θ L (rω)

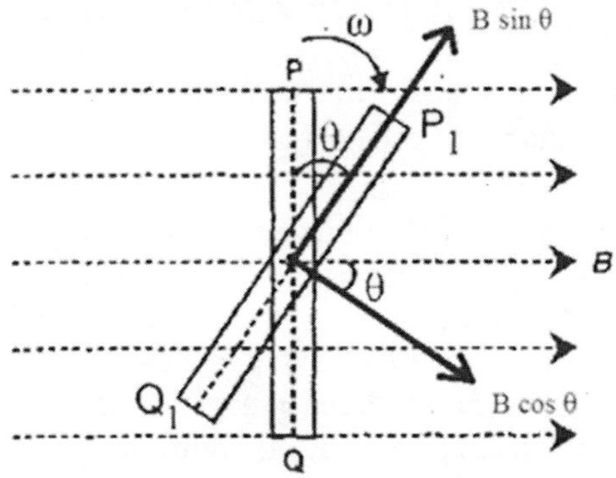

Fig.2.2: Plane of the coil at θ when in the magnetic field

For T turns in one phase

Emf (θ) = 2 B sin θ L (rω) T

Emf (t) = 2 B sin (ωt) L (rω) T

This emf is also an alternating one.

Flux/pole Φ = 2BrL

Total flux = 4BrL

Emf (t) for a single coil = Φω sin (ωt)

Emf (t) per phase = Φω T sin (ωt)

Maximum induced emf/phase = Φ (2πf) T

RMS value of induced emf/phase = Maximum/√2 = 4.44 Φ f T

Here emf equation resembles that in a transformer and hence, the induction motor can be called as the rotating transformer.

If the windings are distributed instead of concentrated, a factor $K_w < 1$ comes into the emf equation. If the pitch of the coil is not full, a pitch factor $K_p < 1$ also need to be included into the emf expression.

RMS value of induced emf/phase in the stator windings= 4.44 K_w K_p Φ f T V

b) Rotor induced EMF

Let the slip be s < 1. Let the mains frequency be *f Hz.*

The mechanical rotation frequency = (1-s) f.

The relative speed of the rotating magnetic field w.r.t rotor windings = sf

The emf induced in the rotor will be at a frequency sf Hz.

The rms value of induced emf in the rotor E_2 = 4.44 Φ sf T Volts = s. E_1

2.3 EQUIVALENT CIRCUIT OF AN INDUCTION MOTOR

Consider a star connected machine with stator resistance be r_1Ω, stator leakage reactance be X_1 Ω, rotor resistance r_2Ω and rotor leakage reactance X_2 Ω.

The induction motor draws negligible current under no load condition from the supply mains just like the transformer and this current sets the rotating magnetic field in the stator. The stator side of the induction motor is represented by Fig.2. 3. The stator circuit draws a no-load current I_o which has two components. One of these components is used to magnetize the stator core (I_m) represented through the inductor with reactance X_m and the other to supply eddy current losses (I_w) in the machine represented through R_c as shown in Fig.2.3.

$$I_o = I_w + I_m$$

In the case of an induction motor, the total magnetising current I_0 is significantly higher, ranging from approximately 25% to 40% of the rated current. This is compared to a transformer, where the magnetising current typically accounts for just about 2% to 5% of the rated current. The higher reluctance observed in the induction motor is due to the presence of an air gap. The magnitude of the magnetising reactance X_0 in an induction motor is also significantly low.

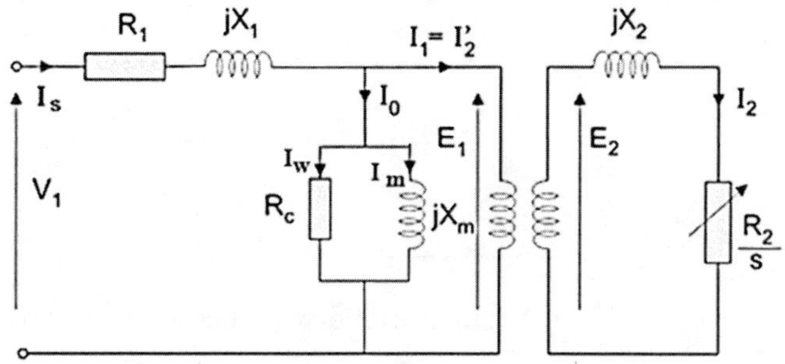

Fig.2.3: The per phase equivalent circuit of IM

When a three-phase supply is given to the three-phase stator winding an emf E_1 is induced in the stator winding. As the rotor is rotating at a speed called slip speed, the emf induced in the rotor is sE_1. The rotor reactance is sX_2. The impedance of the rotor is R_2+jsX_2.

Rotor current $I_2 = sE_1/s\ \surd(R_2/s)^2 + X_2^2) = E_1/\ \surd(R_2/s)^2 + X_2^2)$

Thus, the impedance of the rotor changes as shown in the equivalent circuit of rotor as shown in Fig.2. 4.

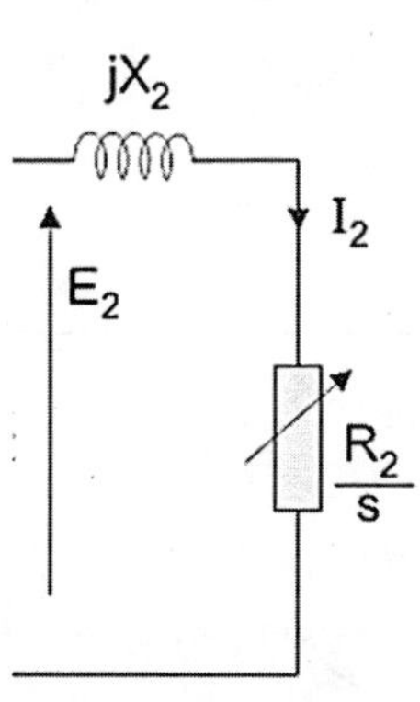

Fig.2.4: Equivalent circuit of rotor

Let us rewrite this impedance further as:

$jX_2 + (R_2/s) = jX_2 + (R_2/s) + R_2 - R_2 = jX_2 + R_2 + (R_2/s)(1\text{-}s)$

This is represented as shown in Fig. 2.5

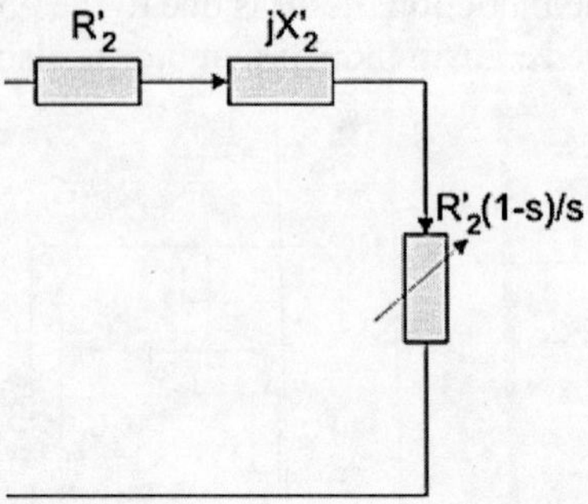

Fig.2.5: Modified Equivalent circuit of the rotor

The quantity $I_2^2R_2$ is the rotor copper loss. The component $I_2^2\,(R_2/s)\,(1\text{-}s)$ is the mechanical equivalent power developed in the rotor.

The equivalent circuit when rotor impedances, voltage, current etc., are shifted to the stator side is as shown in Fig.2.6.

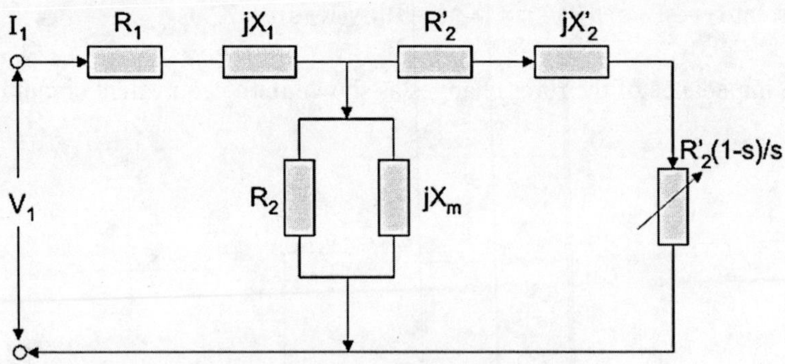

Fig.2.6: Equivalent circuit w.r.t stator

a) Torque of a 3 phase Induction motor

The power output $= 3\,I_2^2\,(R_2/s)\,(1\text{-}s) = \tau\omega_r = \tau\omega_s\,(1\text{-}s)$

Torque developed $\tau = 3\,I_2^2\,(R_2/s)\,(1\text{-}s)\,/\omega_s\,(1\text{-}s) = 3\,I_2^2\,(R_2/s)\,/\omega_s$

The power $3\,I_2^2\,(R_2/s)$ known as the torque in Synchronous Watts.

Torque in Synchronous Watts/ Synchronous speed = True torque developed

b) How to calculate Maximum Torque (Break down Torque)

Torque developed $\tau = 3\, I_2^2\, (R_2/s)\, /\omega_s$
From the rotor side of the equivalent circuit,
$I_2 = sE_2 / \sqrt{(R_2/s)^2 + X_2^2)}$
Therefore, $\tau = ksE_2^2R_2 / (R_2^2 + (sX_2)^2)$

The rotor resistance, rotor inductive reactance, and synchronous speed of an induction motor are all considered constants in this equation. Because the rated voltage that is supplied to the 3-phase induction motor continues to be maintained at a constant level, the stator emf also remains unchanged. The ratio of the emf generated by the rotor to the emf generated by the stator is the definition of the transformation ratio. Therefore, if the emf of the stator is constant, the emf of the rotor must also be constant.

Hence, to find the maximum torque differentiate this w.r.t slip and then equate to zero as all other parameters except slip in the torque equation are constants.

Applying $d\tau/ds = 0$, $s^2 = R_2^2/ X_2^2$

i.e. Slip at which maximum torque occurs (s_m) is obtained as $s = \pm R_2/ X_2$

Neglecting negative value of slip

$s_m = R_2/ X_2$

At starting s = 1, so the maximum torque occurs when rotor resistance = rotor reactance.

Maximum Torque developed $\tau_{max} = kE_2^2 / 2X_2$ N-m.

From the equation mentioned above, it can be concluded that

- The maximum torque is directly proportional to the square of the rotor-induced emf when the rotor is at a standstill.
- The relationship between the maximum torque and rotor reactance is inversely proportional.
- The maximum torque remains unaffected by changes in rotor resistance.

The slip at which the maximum torque is generated is dependent upon the resistance of the rotor, denoted as R_2. By manipulating the rotor resistance, it is possible to achieve maximum torque at any desired slip value. The value of this maximum torque increases with increase in value of R_2.

2.4 TORQUE-SLIP CHARACTERISTICS

a) Low slip/high speed region:

- Let x axis represent the speed and y axis represent the torque. Now, at standstill the speed of the motor is zero at which s = 1 and at synchronous speed s = 0 as shown in the Fig.2. 7. Hence, s is between 0 and 1, it gradually grows from 0 to 1.

- The torque of the motor is given by, $T = ksE_2^2R_2 / (R_2^2 + (sX_2)^2)$

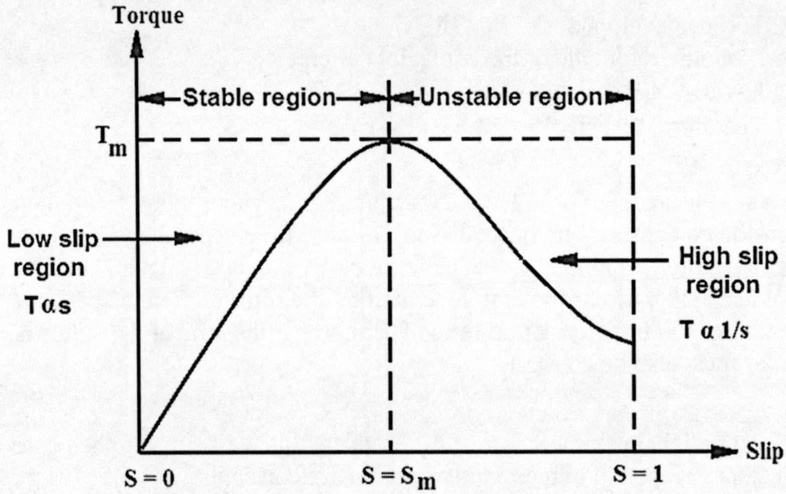

Fig.2.7: Torque – speed Characteristics of Induction Motor

- When the speed of the rotor is close to the synchronous speed, resulting in a low slip, the term $(sX_2)^2$ in the torque equation becomes significantly smaller compared to R_2^2. Hence, there exists a direct relationship between T and s. Therefore, it can be observed that at lower slip values, the torque of an induction motor exhibits a proportional relationship with slip. This relationship is depicted by a linear graph, as illustrated in Figure 2.7.

- The portion of the motor's operation range, specifically from s = 0 to s = sm, is commonly referred to as the low slip or stable region. The operating point of the motor is situated within this region.

- Torque at synchronous speed = 0. At synchronous speed, s = 0 i.e., $R_2/s = \infty$ i.e., Rotor current $I_2 = 0$. Hence no power output and no torque.

- The torque experiences an increase as the slip increases, reaching its maximum value when the slip is equal to R_2/X_2. The maximum value of torque is commonly referred to as the breakdown or pull-out torque. In typical operational scenarios, the pull-out torque of an induction motor is observed to be approximately two to three times greater than the rated full-load torque. Hence, the motor is capable of enduring short-term overloads without stalling.

b) High slip/low speed region:

- When the load exceeds the maximum torque, the slip is increased, resulting in a significantly larger value for the term $(sX_2)^2$ compared to R_2^2. Hence, the effect of R_2^2 is ignored incomparison with $(sX_2)^2$, resulting in an updated expression for the torque.

$T = ksE_2^2R_2 / (sX_2)^2$ or $T = kE_2^2R_2 / (sX_2^2)$

In the above expression, all variables remain constant except for the value of s.

Therefore, T α 1/s. As a result, at higher slip values, torque is inversely proportional to slip, s, and the induction motor's torque slip characteristics are rectangular hyperbolas as shown in Fig.2. With an increase in load in this region, slip increases but torque decreases. As a result, the motor fails to carry the load, slows down, and finally stops. This high slip region (between $s = s_m$ and s = 1) is referred to as an unstable region as a result.

- The locked rotor torque, also known as the starting torque, is the torque that the motor produces when it starts at zero speed or at rest, as shown in Fig.2.8. For starting heavy or hard types of loads like positive displacement pumps, cranes, traction, etc., a high starting torque is particularly important. In situations where the required starting torque is low or close to zero, such as centrifugal fans or pumps, a lower starting torque can be accepted. At the start, slip equals 1. The starting torque is the corresponding torque.
  $\tau_{stg} = 3\ E_1^2\ R_2 / (R_1+R_2)^2 + X^2)\ \omega_s$

- The pull-up torque is the minimum torque in Figure 2.8 and is produced by the motor during its acceleration from zero to full-load speed, prior to reaching the breakdown torque threshold. When the motor initiates and begins acceleration, the torque typically undergoes a gradual reduction until it achieves a minimum value at a specific speed, known as the pull-up torque point. Subsequently, the torque experiences an increase until it reaches its maximum value at a higher speed, referred to as the break-down torque point. The pull-up torque is of utmost importance in applications that require the transmission of power via temporary barriers in order to achieve the desired working conditions.

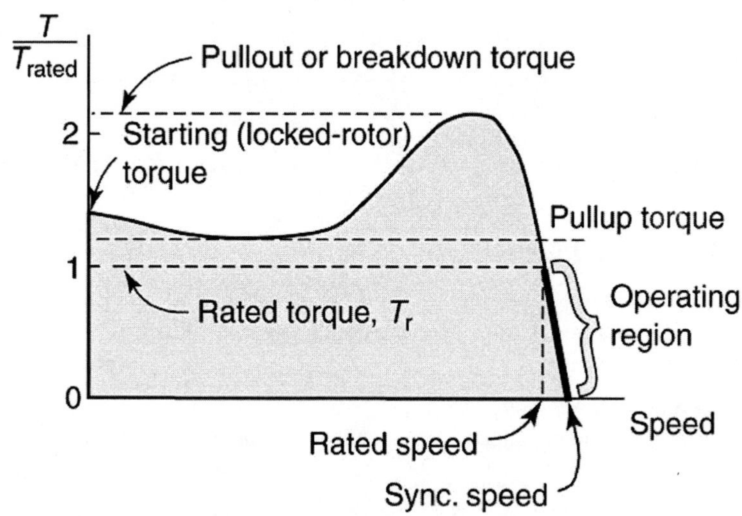

Fig.2.8: Various torques marked in the Torque – speed Characteristics of Induction Motor

- Full-load torque refers to the torque necessary to produce the specified power output of an electric motor when operating at its maximum speed under full-load conditions. Figure 2.9 illustrates a normal torque-speed curve that displays two different loads exhibiting an identical steady running speed (N). The torque-speed curve of the motor is shown by the solid line, whereas the load characteristics are represented by the two dotted lines. Load (A) exhibits characteristics commonly found in a basic hoist, wherein the motor experiences a constant torque regardless of its rotational speed. Conversely, load (B) can be likened to that of a fan. The term "T_{acc}" denotes the accelerating torque, which signifies the disparity between the torque generated by the motor and the torque necessary to operate the load at a given velocity.

- The operational torque and speed of the motor are determined by the intersection of these properties. To ensure the selection of the appropriate motor for this load, it is important to accurately determine the torque-speed characteristic of the driven load.

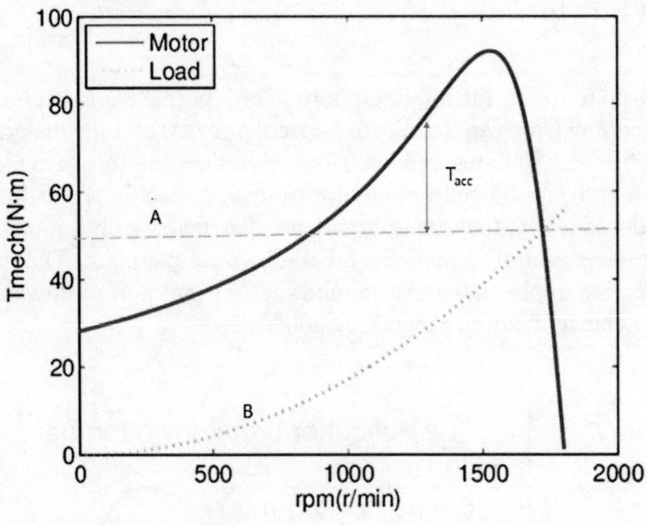

Fig.2.9: Motor Torque – speed Characteristics with that of the load requirement

2.5 INDUCTION GENERATOR

The induction machine can be used as a generator. It is alternatively referred to as an asynchronous generator. Under specific conditions, an induction machine has the capability of acting as a generator.

a) Slip becomes negative.
b) The electric torque produced is greater than the prime mover torque.

Let us consider an induction machine that is connected to a prime mover with controllable speed. If the speed of the prime mover is increased to the point where the slip becomes negative, meaning that the speed of the prime mover exceeds the synchronous speed, If the speed of the prime mover is increased beyond the negative maximum torque value, the generator's generating effect ceases to exist. The induction generator's speed throughout its operation is evidently not synchronous; hence, it is called an asynchronous generator.

Since the induction generator is not a self-excited machine, reactive power and magnetising current are needed in order to create a rotating magnetic field. The supply mains or another synchronous generator is some of the sources the induction generator uses to get its magnetising current and reactive power. Since the induction generator constantly needs reactive power from the supply system, it cannot operate alone. An isolated or self-excited induction generator, on the other hand, might use a capacitor bank for reactive power supply rather than an AC supply system. Let's think about the isolated induction generator shown in Figure 2.10. When an induction generator is connected to the grid or a stand-alone system, the reactive power needed to create the air gap magnetic flux is supplied by a capacitor bank connected to the machine, while in the latter case, the reactive power is drawn from the grid.

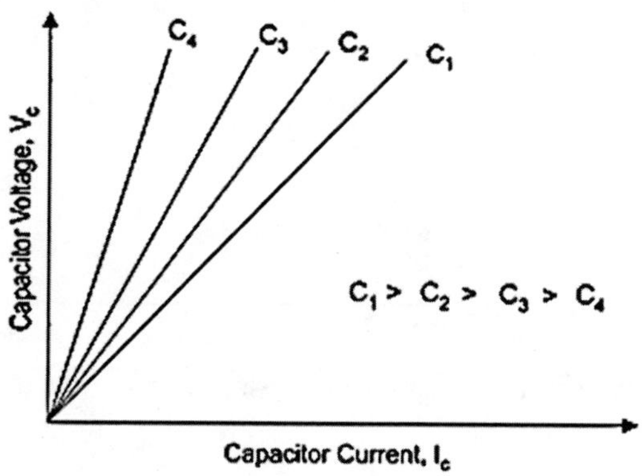

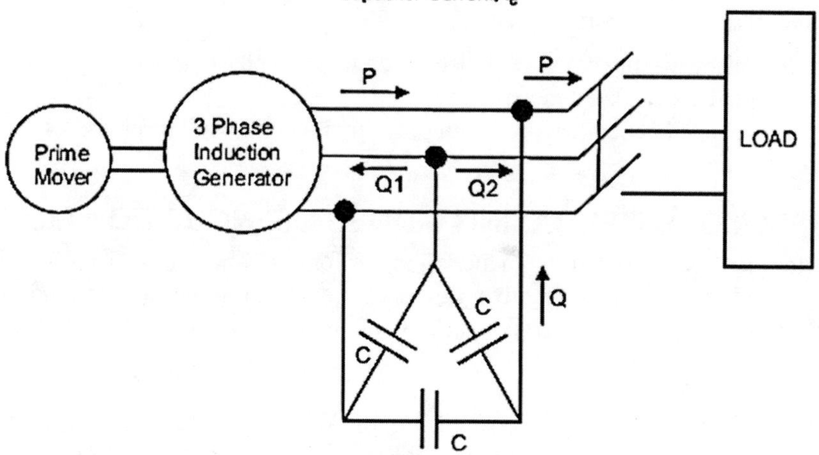

Fig.2.10: Self-excited Induction generator

The induction generator and load are both provided with lagged reactive power by the capacitor bank in this situation. i.e., the induction generator's and the load's combined reactive power consumption is equal to the capacitor bank's total reactive power supply. When the rotor of the induction machine rotates at the proper speed, there is the formation of a small terminal voltage *oa* across the stator terminal, as shown in Figure 2.11. This voltage *oa* causes the capacitor current *ob* to be produced. The voltage de is produced by the current *bc* sending current to *od*. Until the induction generator's saturation curve crosses the capacitor load line at some point, the cumulative process of voltage creation continues. In the provided curve, this point is known as f.

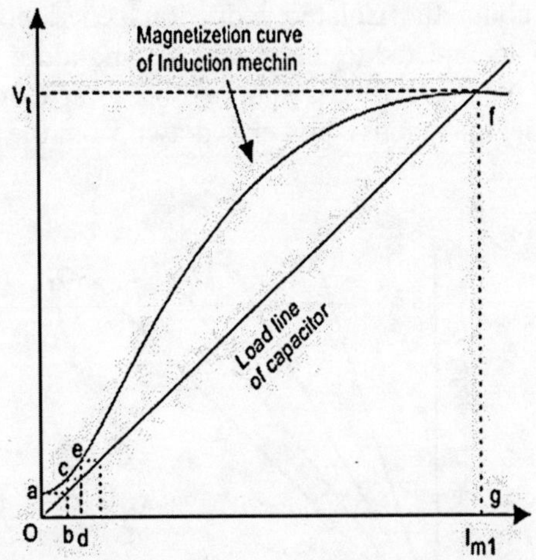

Fig.2.11: Generated voltage vs Magnetization current of the induction generator

Application of Induction Generator

- Regenerative breaking of hoists driven by 3-phase induction motors frequently uses externally stimulated generators.
- Self-excited generators are commonly employed in wind turbines.

2.6 VARIOUS OPERATING MODES OF INDUCTION MOTOR

The torque-speed curve of the induction motor for constant rotor resistance showing variation of torque with positive and negative values of s demonstrating various modes of operation like Motoring region, generating region and Breaking region as shown in Fig.2.12.

a) Motoring Region

The slip in motoring mode is between 0 and 1. When a 3-phase supply is applied to the stator, a revolving flux rotating at synchronous speed is formed, and the rotor maintains a relative speed known as the slip speed to facilitate unidirectional torque. At synchronous speed, the slip is zero, while at standstill, the slip is one. As slip goes from 0 to 1, the torque of the motor changes from zero to full-load torque. This is the induction motor's general operating range.

b) Generating Region

During the generating mode, the induction motor operates at a speed higher than the synchronous speed, exhibiting the characteristics of an induction generator. In order to achieve a speed greater than the synchronous speed, an external prime mover is used. Therefore, both slip and torque exhibit negative values. In this operational mode, the machine undergoes the conversion of mechanical energy into electrical energy. In the event of a stand-alone system, the establishment of the air gap magnetic flux necessitates the provision of reactive power through a capacitor bank linked to the machine. Conversely, in the case of grid connections, the maintenance of the air gap flux requires the drawing of reactive power from the grid.

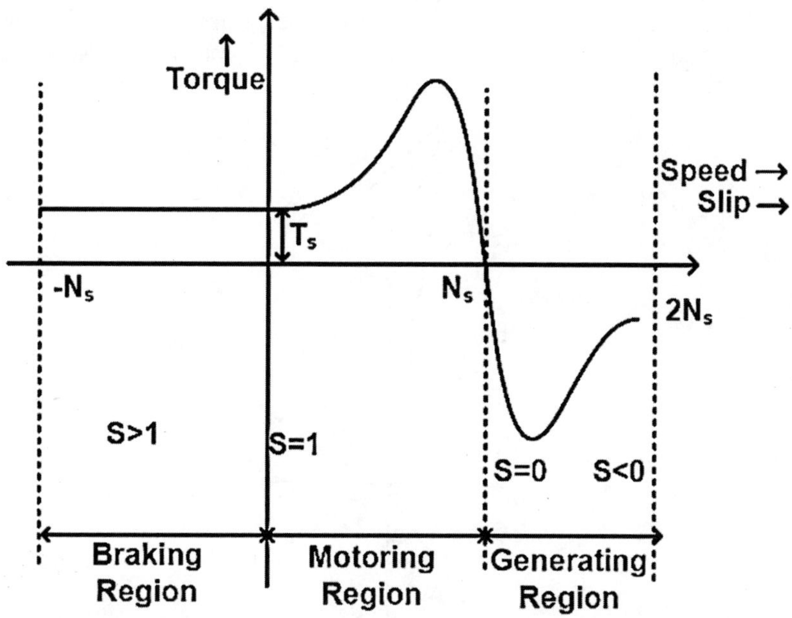

Fig.2.12: Torque-Speed Characteristics of Induction machine under various modes of operation

c) Braking Region

In this braking region, the polarity of the supply voltage is reversed, causing the motor to reverse its direction and reverse the current. This type of electrical braking is referred to as plugging. The slip is more than one when braking. Heat is produced as a result of the kinetic energy stored in the load. It is also created as heat if the stator is connected to the power supply. As a result, before entering braking mode, the supply from the stator must be disconnected. This procedure causes the motor to stop in a short period of time.

2.7 PHASOR DIAGRAM OF INDUCTION MOTOR

A phasor diagram is a graphical representation that illustrates the phase relationship between the voltages and currents in an electrical system as shown in Fig.2.13. In the case of an induction motor, the phasor diagram provides a visual representation of the electrical quantities and their relationships, aiding in the understanding of the motor's operation, power flow, and performance characteristics.

Stator Current: The stator current is represented by a phasor, typically denoted as I_1. It is in phase with the stator voltage (V_1) in a balanced 3-phase system.

Rotor Current: The rotor current is typically denoted as I_2. It lags behind the stator current by an angle (θ_2) known as the rotor impedance angle. The rotor current is induced by the rotating magnetic field of the stator.

Stator Voltage: The stator voltage, denoted as V_1, is the applied 3-phase voltage across the stator winding. It is considered as the reference phasor and is typically represented along the horizontal axis.

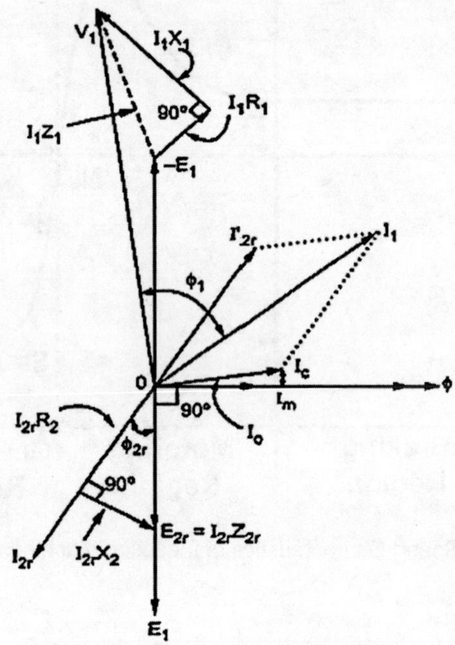

Fig.2.13: Phasor diagram of the induction motor

Rotor Voltage: The rotor voltage, denoted as V_2, is induced by the relative motion between the rotating magnetic field and the rotor conductors. It lags behind the rotor current by an angle (θ_2) due to the rotor impedance.

Power Factor: The power factor in an induction motor is the cosine of the phase angle between the stator current (I_1) and the stator voltage (V_1). A lagging power factor indicates an inductive load.

Slip: Slip is the relative speed difference between the rotating magnetic field and the rotor. It is denoted as *s* and can be represented as a ratio or percentage. Slip is directly related to the rotor current and torque production.

Torque Angle: The torque angle, denoted as δ, represents the phase angle difference between the stator current (I_1) and the rotor current (I_2). It determines the torque production and power transfer in the motor.

The phasor diagram of an induction motor operating under load conditions exhibits similarities to that of a loaded transformer. The main difference lies in the operational characteristics of the secondary in an induction motor and a transformer. In an induction motor, the secondary is both rotating and short-circuiting, while in a transformer, the secondary remains stationary and is connected to a load. The load applied to an induction motor is mechanical, while the load imposed on a transformer is electrical. The phasor diagram of an induction motor can be created by determining the electrical equivalent of the mechanical load on the motor.

Let Φ represent the magnetic flux that is associated with both the primary and secondary sides. E_1 denotes the self-induced e.m.f. in the stator, while E_{2r} represents the mutually induced e.m.f. in the rotor. Let R_1 denote the resistance per phase of the stator, and X_1 represent the reactance per phase of the stator.

The voltage per phase V_1 of the stator must be sufficient to counteract the self-induced e.m.f. E_1 and provide for the voltage drops $I_1 R_1$ and $I_1 X_1$. The voltage applied to the stator,

$$\bar{\bar{V}}_1 = -\bar{\bar{E}}_1 + \overline{\overline{I_1 R_1}} + j\overline{\overline{I_1 X_1}} = -\bar{\bar{E}}_1 + \overline{\overline{I_1 Z_1}}$$

In a running condition, the rotor-induced e.m.f. has to supply the drop across impedances due to the rotor being short-circuited, and this is given by:

$$\bar{\bar{E}}_{2r} = \overline{\overline{I_{2r} R_2}} + j\overline{\overline{I_{2r} X_2}} = \overline{\overline{I_{2r} Z_{2r}}}$$

E_{2r} is proportional to the number of rotor turns relative to the number of stator turns. I_{2r}, the rotor current when running, lags E_{2r} by a rotor p.f. angle of Φ_{2r}.

The effect of load is represented by the reflected rotor current I_{2r}' on the stator side and is given by:

$$I_{2r}' = K\, I_{2r}$$

The induction motor draws current I_o, the phasor sum of I_c and I_m, while it is not under load. The total stator current taken from the source is:

$$\bar{I}_1 = \bar{I}_o + \bar{I}_{2r}'$$

The induction motor's power factor is determined by the angle of Φ_1, or the cosine of Φ_1, between V_1 and I_1.The induction motor's phasor diagram can be produced using all of the relationships mentioned above.

To draw a phasor diagram, follow these steps:

1. Use Φ the phasor as a reference.
2. There is a 90º lag in the induced voltage E_1.
3. To illustrate - E_1, reverse the voltage phasor E_1. In phase with E_1 is the phasor E_{2r}. Draw I_{2r}, which is lagging E_{2r}, or the E_1 direction, at an angle of Φ_{2r}.
4. To determine the precise location of E_{2r}, display I_{2r} R_2 drop in phase with I_{2r} and I_{2r} X_{2r} leading the resistive drop by 90º.
5. To get I_{2r}', reverse I_{2r}.
6. I_c is in charge while I_m is in phase with. To get I_o, combine I_m and I_c.
7. To get I_1, add I_o and I_{2r}'.
8. Add I_1 R drop in phase with I_1 and I_1X_1 at 90 degrees, starting at the tip of the E_1 phasor,to get I_1 to V_1 phasor.
9. There is a Φ angle between V_1 and I_1.

2.8 POWER FACTOR IN INDUCTION MOTORS

Power factor is defined as the ratio of the real power (active power) to the apparent power consumed by the motor. It is a measure of the efficiency of power utilization by the motor and is an important parameter in assessing the electrical system's overall performance. In an induction motor, the power factor is influenced by the characteristics of the motor itself, the load it is driving, and the operating conditions.

The power factor in an induction motor is primarily determined by two factors:

Magnetizing (or Core) Power: Induction motors require a certain amount of reactive power to establish the magnetic field in the motor's stator. This reactive power, known as magnetizing power, is necessary for the motor to operate but does not contribute to useful work. It creates an inductive component in the motor's current, leading to a lagging power factor.

Load Power: The active power (real power) consumed by the motor to perform mechanical work is known as the load power. It contributes to useful work, such as driving a pump or rotating a fan. The load power determines the overall power factor and can be either leading or lagging, depending on the nature of the load. The power factor in an induction motor is typically lagging (inductive) and is denoted by a lagging angle (θ) between the stator current and voltage. A lagging power factor indicates that the motor is drawing more reactive power than necessary, which can result in decreased system efficiency and increased line losses.

Improving the power factor of an induction motor can be achieved through various methods, including:

i) **Power Factor Correction Capacitors:** Installing power factor correction capacitors in parallel with the motor helps compensate for the reactive power requirements, reducing the lagging power factor and improving overall system efficiency.

ii) **Variable Frequency Drives (VFDs):** Using VFDs allows for speed control of the motor and can provide improved power factor performance by adjusting the motor's voltage and frequency.

iii) **Proper Sizing and Selection:** Selecting an induction motor with a power rating appropriate for the load requirements can help optimize power factor performance.

Induction motors benefit from maintaining a high power factor since doing so lowers energy losses, boosts system effectiveness, and puts less strain on the electrical distribution network. When developing, running, and maintaining induction motor systems, it's crucial to take power factor into account in order to maximise energy efficiency and overall system performance.

A lagging (inductive) power factor is presented to the power line by induction motors. For big high-speed motors that are completely loaded, the power factor can be as favourable as 90%. The highest high-speed motor power factor is 92% at 3/4 load. Small low-speed motors can have power factors as low as 50%.

As the rotor increases speed, the power factor increases, rising from 10% to 25% at start-up. The motor mechanical load has a significant impact on power factor (PF) (as is evident from Fig. 2.14.). Similar to a transformer with no resistive load on the secondary is an unloaded motor. The secondary (rotor) to the primary (stator) reflect very little resistance. Thus, a reactive load as low as 10% PF is observed on the power line. The power factor rises as a result of an increased resistive component being reflected from the rotor to stator when the rotor is loaded.

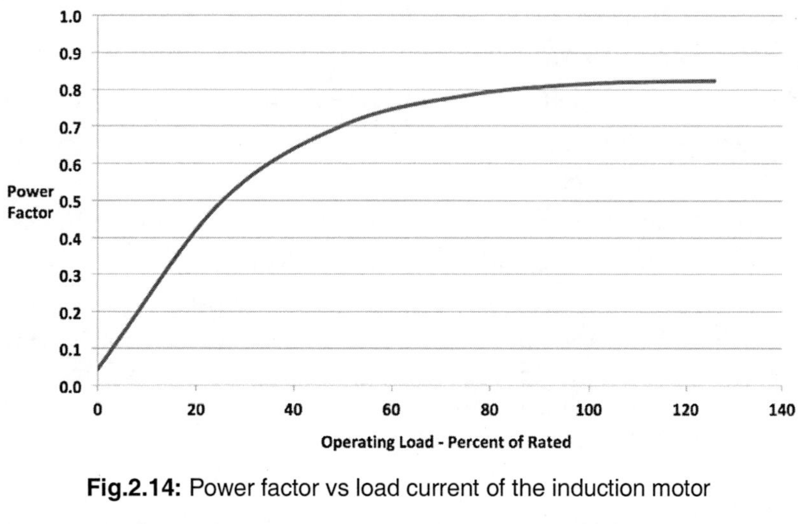

Fig.2.14: Power factor vs load current of the induction motor

2.9 POWER FLOW DIAGRAM AND LOSSES OF INDUCTION MOTOR

Power Flow Diagram of Induction Motor explains the input given to the motor, the losses occurring, and the output of the motor. The input power given to an induction motor is in the form of three-phase voltage and currents.

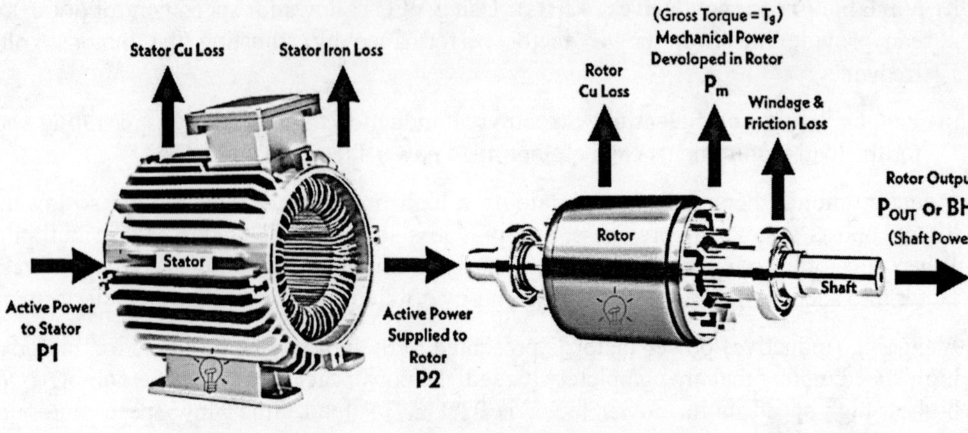

Fig.2.15: Power flow diagram and losses of Induction motor

The Power Flow Diagram of an Induction Motor is as shown in Fig.2.15.

Input power to stator $P_1 = 3\ V_1 I_1 \cos(\theta)$.
where, V_1 is the stator voltage applied.
I_1 is the current drawn by the stator winding.
$\cos(\theta)$ is the stator power factor.

Losses in an Induction Motor:

The following list of losses that happen during energy conversions in an induction motor includes: (i) Constant losses (ii) Variable losses.

(i) Constant losses: Constant losses refer to losses that are unaffected by changes in the load and remain constant despite load variations. These losses may be:

(a) Core losses: The losses observed in both the stator and rotor cores include hysteresis and eddy current effects. The losses due to eddy currents in the rotor core can be considered small, as the frequency of the rotor current is quite low, typically ranging from 0.5 to 2 Hz. The mentioned losses can be classified as constant or fixed losses, as they depend on the voltage and frequency, both of which are typically maintained at a constant level.

(b)Friction and windage losses: These losses remain constant as they are directly influenced by the speed of the induction motor. The speed of an induction motor remains nearly constant during normal running, with a very small slip. These losses occur in the machine as a result of power being used to overcome friction at the bearings and wind resistance. There may be some sliding friction loss that occurs in the slip ring induction motor.

(ii) Variable losses: Variable losses are losses that depend on the load and change as the load changes. These losses are:

(a) I^2R loss in stator winding.

(b) I^2 R loss in rotor winding.

These losses are also constant because they are proportional to the speed of the induction motor. The speed of an induction motor is nearly constant (slip is very small while running normally). These losses occur in the machine as a result of power loss owing to bearing friction and to overcome wind resistance. The slide ring induction motor suffers from additional sliding friction loss.

(c) Brush contact loss: Only slip-ring induction motors have this type of loss. This occurs because brushes and slip rings have contact resistance. Its magnitude is very small because its contact resistance is kept to a minimum.

(iii) Stray losses: These losses take place in both the iron core and the winding of the machine. These cannot be determined exactly but are accounted for, but they are taken intoconsideration during the calculation of the machine's efficiency by a suitable factor.

The **losses** in the stator are: $\mathbf{P_s = 3\ I^2R_s + P_h + P_e}$

The output power of the stator is given as: $\mathbf{P_{os} = P_{is} - (3\ I^2R_s + P_{hs} + P_{es})}$

The power produced by the stator passes to the rotor of the machine through the air gap between the stator and the rotor. The power of the machine is referred to as the **air gap power**, denoted as Pg.

Thus, *Power output of the stator = air gap power = input power to the rotor*

Rotor input = Power input - Stator copper and iron losses.

The losses in the rotor are as follows:

I^2R losses in the rotor resistance.

They are also called **Rotor copper losses** and represented as: $\mathbf{P_{rc} = 3\ I^2R_r}$

Hysteresis and eddy current losses in the rotor core. They are known as **Rotor core losses.**

Rotor Copper loss = Slip × power input to the rotor.

Developed Power = (1 – s) × Rotor input power.

Friction and Windage losses P_{fw}

Stray load losses P_{misc}, which includes all losses not covered elsewhere, such as losses caused by harmonic fields. When the copper losses in the rotor are subtracted from the input power of the rotor (P_g), the resulting power represents the conversion of electrical energy into mechanical energy. The term used to describe this concept is "Developed Mechanical Power P_{md}".

Developed Mechanical power = Rotor input – Rotor copper loss

The equation provided represents the output of the motor:

$\mathbf{P_o = P_{md} - P_{fw} - P_{misc}}$

P_o is called the **shaft power** or the **useful power.**

Rotational losses

The rotor core losses are high during acceleration and at start up. These losses reduce as the induction motor's speed increases. At first, there are no losses due to friction or windage. The losses start to rise as the speed does as well. With a change in speed, the total losses from friction, windage, and core remain nearly constant. These losses are collectively referred to as **rotational losses.**

It is given by the equation shown below:

$P_{rot} = P_{fw} + P_h + P_e + P_{misc}$

$P_o = P_{md} - P_{rot}$

The rotational losses are not explicitly included in the equivalent circuit, but they are purely related to mechanical aspects.

Efficiency

The standard definition of efficiency, as denoted by the ratio of output power to input power, when applied to the induction motor.

$$\eta_m = \frac{\textit{Mechanical power output in watts}}{\textit{Mechanical power output in watts} + \textit{Fixed losses in watts} + \textit{Copper losses in watts}} \times 100$$

or $$\eta_m = \frac{P_{mech}}{P_{mech} + P_{const} + P_{cu}} \times 100$$

2.10 FOUR QUADRANT OPERATION OF MOTOR DRIVE

The torque and speed of the motor are mechanically related to the electrical quantities of current and voltage. Let's look at the conventions that determine the four-quadrant operation of motor drives. A positive motor speed is associated with rotation in the forward direction. The forward speed of drives that operate in only one direction will be their normal speed. When dealing with loads that move in both upward and downward directions, the motor's speed responsible for the upward motion is typically referred to as the forward motion speed. In the case of reversible drives, the forward speed is selected without any specific criteria or constraints. A negative sign denotes the reverse speed, which is associated with rotation in the opposite direction. Positive motor torque refers to the torque that generates acceleration or the rate of increase in speed in the forward direction. The motor torque is determined by

$$T = T_l + J\frac{d\omega_m}{dt}$$

The direction of positive load torque is the exact opposite of positive motor torque. If a motor produces deceleration, it is considered to have negative torque.

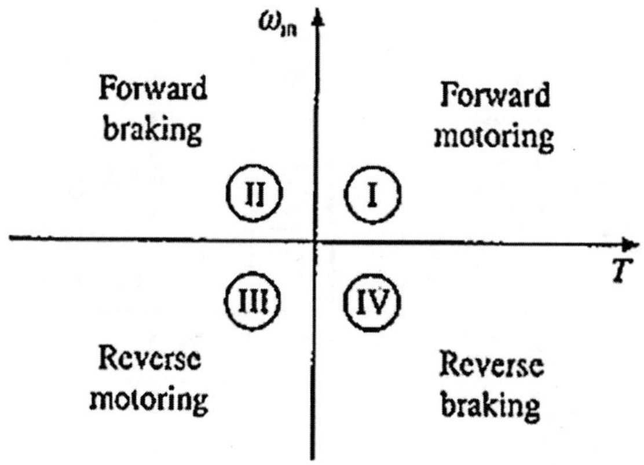

Fig.2.16: Four Quadrant operation of the drive.

There are two modes in which a motor operates: motoring and braking. It moves by converting electrical energy into mechanical energy through motoring. It acts as a generator when braking, transforming mechanical energy into electrical energy and blocking motion. Both forward and reverse motoring and braking functions are possible with motors.

The Four Quadrants Operation of the motor drive is shown in Figure 2.16 with the torque and speed coordinates for both forward (positive) and reverse (negative) motions. The ratio of a motor's speed to torque determines its output power. The developed power is positive in quadrant I. As a result, the machine functions as a motor to supply mechanical energy. Therefore, operation in quadrant I is known as forward motoring. Power is negative in quadrant II. As a result, the machine operates by braking against its forward motion. Thus, forward braking is the activity of quadrant II. In a similar manner, operations in quadrants III and IV can be distinguished as braking and reverse motoring, respectively.

Let's look at how a hoist operates in the Four Quadrant Operation of Motor Drive in Fig.2.17 to get a better understanding of the notations above. Arrows indicate the motor and load torque directions as well as the speed direction.

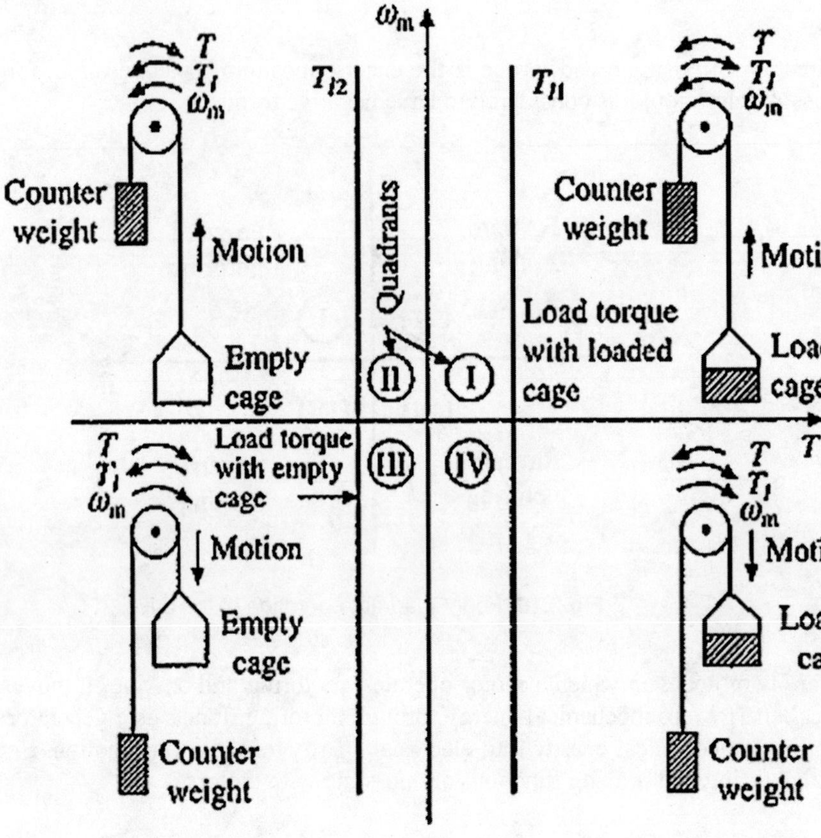

Fig.2.17: Four Quadrant operation of a motor driving a hoist load.

A rope wound on a drum attached to the motor shaft makes up a hoist. The cage, which is used to move people or objects from one level to another, is attached to the rope's one end. A counterweight is attached to the rope's other end. The counterweight's weight is chosen to be higher than the weight of a cage that is completely empty but lower than the weight of a cage that is completely loaded. The cage will go higher when the motor is moving in the forward direction. Fig.2.17 also displays the speed-torque parameters of the hoist load. It is useful to plot the positive load torque on the same axis as the positive motor torque, even though Eq. (2.2) states that they are opposite in sign. In fact, the load-torque curve represented in this way is the opposite of the actual Load torque remains constant and unaffected by speed. This is mostly true with a low-speed hoist, where forces due to friction and windage are so small that they can be considered negligible compared to those due to gravity. Gravitational torque remains consistent and unaffected, regardless of whether the direction of the driving motor is reversed. The load torque line Tl1 in quadrants I and IV represents the positive speed-torque characteristic of the loaded hoist. This torque represents the potential for increased efficiency and balance between the loaded hoist and counterweight. The load torque line Tl2 in quadrants II and III represents the speed-torque characteristic of a hoist that is ready for new challenges! This torque is the positive impact of the counterweight on the hoist. Its sign is positive because the weight of a counterweight is always higher than that of an empty cage.

The quadrant I operation of a hoist allows for the uplifting movement of the cage, which corresponds to the positive motor speed, moving in an anticlockwise direction. This motion will be achieved when the motor generates a positive torque in an anticlockwise direction, matching the magnitude of the load torque Tl1. Since developed motor power is positive, this is an exciting forward motoring operation!

Quadrant IV operation is achieved when a loaded cage is safely lowered. The weight of a loaded cage is higher than that of a counterweight, which means it has the power to come down effortlessly thanks to gravity. To ensure the cage's speed is within a safe range, the motor will generate a positive torque T, equal to Tl2, in an anticlockwise direction. This will help maintain a secure environment. As both power and speed are negative, drive is operating in reverse braking, but this means we have the opportunity to slow down and make a controlled stop.

Operation in quadrant II is achieved when an empty cage is moved upwards. Because a counterweight is heavier than an empty cage, it has the strength to effortlessly lift it up. To ensure safety, the motor will generate a braking torque equal to Tl2 in a clockwise (negative) direction, effectively limiting the speed to a safe value. Since speed is positive and developed power is negative, it is an opportunity for controlled deceleration.

Operation in quadrant III is achieved when a cage is lowered and ready to be filled with new possibilities. Since an empty cage has a lighter weight than a counterweight, the motor can easily produce torque in a clockwise direction. Since speed is negative and power is positive, this is an exciting reverse motoring operation.

UNIT SUMMARY

The study of characteristics of torque versus slip (speed) in an induction motor reveals essential insights into its performance under various conditions. These characteristics include starting torque, full load torque, and maximum torque, each of which plays a role in understanding the motor's capabilities and limitations. Viewing the induction motor as a generalized transformer with phasor diagrams simplifies the analysis of its electrical and magnetic aspects, aiding in comprehending its behaviour and facilitating calculations.

Four quadrant operation is a crucial concept that allows the motor to operate in both forward and reverse directions, enabling versatile motor control and applications that require frequent changes in direction, like electric vehicles and industrial machinery. The power flow diagram provides a visual representation of energy transfer between the electrical power source and the mechanical load, helping optimize motor efficiency and identify potential energy losses.

EXERCISES

1. Plot the torque-speed characteristics for a typical 3-phase induction motor on a graph. Label and explain the regions corresponding to starting, accelerating, full-load, and maximum torque.

2. Draw a phasor diagram representing the primary and secondary voltage and current vectors of a 3-phase induction motor operating at a given load condition. Include the rotor impedance angle and explain how it affects the diagram.

3. Given the primary voltage and current phasors of an induction motor, calculate the secondary (rotor) current phasor and angle. Use this information to determine the motor's power factor.

4. Describe the concept of four-quadrant operation in the context of an induction motor. Explain the significance of positive and negative torque and positive and negative slip.

5. Create a power flow diagram for a regenerative braking system using a four-quadrant-capable induction motor. Show the power flow between the motor, the load, and the electrical grid in both motoring and regenerating modes.

6. Given the mechanical and electrical power input and output values for an induction motor, calculate its efficiency for different load conditions. Illustrate these efficiency values on a power flow diagram.

Multiple Choice Questions

1. Which characteristics of an induction motor provide insights into its performance under various conditions?

   a) Slip-Speed Characteristics

   b) Power Flow Characteristics

   c) Torque-Speed Characteristics

   d) Voltage-Current Characteristics

2. Starting torque of an induction motor is crucial because it determines:
   a) The motor's maximum speed.
   b) The motor's full load torque.
   c) Whether the motor can overcome initial resistance and start moving.
   d) The motor's efficiency at rated load.

3. Which torque of an induction motor is used to assess its efficiency under normal operating conditions?
   a) Maximum Torque
   b) Starting Torque
   c) Full Load Torque
   d) Stall Torque

4. The induction motor can be represented as a generalized transformer using:
   a) Voltage-Angle Diagram
   b) Phasor Diagram
   c) Power Flow Diagram
   d) Slip-Speed Diagram

5. Four quadrant operation of an induction motor refers to its ability to:
   a) Operate in four different voltage quadrants.
   b) Operate at four different speeds.
   c) Operate in both forward and reverse directions.
   d) Generate power in four distinct power ranges.

6. The power flow diagram of an induction motor helps in:
   a) Selecting the appropriate starter for the motor.
   b) Analyzing the slip-speed characteristics.
   c) Optimizing motor efficiency and identifying potential energy losses.
   d) Understanding the motor's maximum speed.

7. Which characteristic is essential for assessing the motor's ability to handle varying loads?
   a) Starting Torque
   b) Full Load Torque

c) Maximum Torque

d) Stall Torque

Answers of Multiple-Choice Questions

1. c. Torque-Speed Characteristics
2. c. Whether the motor can overcome initial resistance and start moving.
3. c. Full Load Torque
4. b. Phasor Diagram
5. c. Operate in both forward and reverse directions.
6. c. Optimizing motor efficiency and identifying potential energy losses.
7. c. Maximum Torque

Short and Long Answer Type Questions

Short answer questions

1. What does the torque-speed characteristic of an induction motor show?
2. Define starting torque, full load torque, and maximum torque in an induction motor.
3. How are starting torque, full load torque, and maximum torque related to each other in an induction motor?
4. How can an induction motor be represented as a generalized transformer?
5. What does a phasor diagram in an induction motor represent?
6. What is the significance of the rotor current in the phasor diagram?
7. What is meant by four quadrant operation in an induction motor?
8. How is a power flow diagram used in an induction motor analysis?
9. What does the power flow diagram indicate about an induction motor?

Long Answer type questions

1. Describe the torque-speed characteristic of an induction motor in detail. Explain how the torque output varies with slip and discuss the significance of starting torque, full load torque, and maximum torque in motor operation.
2. How are starting torque, full load torque, and maximum torque related to each other in an induction motor? Explain the concept of breakdown slip and its importance in motor performance.
3. Elaborate on the practical applications of the torque-speed characteristic curve in selecting and controlling induction motors for different industrial and commercial applications.
4. Explain the concept of representing an induction motor as a generalized transformer. How does the stator winding function as the primary winding, and what is the role of the rotor in the analogy?
5. Draw a phasor diagram for an induction motor and explain the significance of stator and rotor currents and voltages. How does the phasor diagram help visualize the electrical and magnetic interactions within the motor?

6. Discuss the key differences between the phasor diagram of an induction motor operating at no-load and full load conditions. How does the power factor and magnetizing current change between these two scenarios?
7. What is four quadrant operation in the context of an induction motor? Explain with examples how the motor can operate in both forward and reverse directions and perform motoring and braking in each quadrant.
8. Construct a power flow diagram for a three-phase induction motor, and explain the significance of active power, reactive power, apparent power, and power losses in the motor system.
9. How does the power flow diagram help identify the efficiency and losses in the motor? Discuss the importance of power factor correction in improving motor efficiency.
10. Discuss the need for using starters in induction motors during start-up. Explain why the starting current of induction motors is much higher than the full load current and how starters help in limiting this current.
11. Compare and contrast various types of motor starters, including stator resistance, auto transformer, star-delta, rotor resistance, and soft starters. Explain the working principles and applications of each starter type.
12. Suppose you need to start a high-power induction motor with limited voltage fluctuations during start-up. Recommend the most suitable starter type for this scenario and justify your choice based on efficiency, torque characteristics, and impact on the power supply.

Numerical Problems

1. An induction motor has the following data:

   Full Load Torque (T_{FL}) = 100 Nm; Maximum Torque (T_{MAX}) = 150 Nm; Synchronous Speed (N_s) = 1500 RPM. Calculate the slip at which the motor produces Full Load Torque and Maximum Torque.

Solution:

Given:

T_{FL} = 100 Nm

T_{MAX} = 150 Nm

N_s = 1500 RPM

Slip $s = (N_s - N_r) / N_s$

where N_r is the motor speed in RPM.

For Full Load Torque: $s_{FL} = (N_s - N_{FL}) / N_s$

For Maximum Torque: $s_{MAX} = (N_s - N_{MAX}) / N_s$

Let's assume:

N_FL = 1440 RPM (Speed at Full Load Torque)

N_MAX = 1200 RPM (Speed at Maximum Torque)

Calculating Slips:

s_{FL} = (1500 - 1440) / 1500 = 0.04 or 4%

s_{MAX} = (1500 - 1200) / 1500 = 0.20 or 20%

Answers:

Slip at Full Load Torque (s_{FL}) = 4%

Slip at Maximum Torque (s_{MAX}) = 20%

2. For an induction motor with stator current (I_s) of 10 A and rotor current (I_r) of 5 A, draw a phasor diagram considering the stator current as the reference.

Solution:

In a phasor diagram, vectors represent the magnitude and phase angle of currents or voltages. Since the stator current (I_s) is considered the reference, its phasor is drawn horizontally (0°).

Answers:

Let's represent stator current (I_s) = 10 A and rotor current (I_r) = 5 A.

The phasor diagram will show the stator current (I_s) as a horizontal line (0°) with a magnitude of 10 A, and the rotor current (I_r) as a vector at an appropriate angle (positive or negative) with a magnitude of 5 A.

3. Calculate the active power (P), reactive power (Q), apparent power (S), and power losses (P_{Loss}) in the three-phase induction motor. Given, the following information:

Line voltage (V_L) = 400 V (rms), Line current (I_L) = 50 A (rms), Power factor (pf) = 0.85 (lagging), Motor efficiency (η) = 0.92

Solution:

The apparent power (S) of the motor

$S = \sqrt{3} * V_L * IL = \sqrt{3} * 400 * 50 \approx 34,641$ VA

The active power (P) of the motor

$P = S * pf = 34,641 * 0.85 \approx 29,445$ W

The reactive power (Q) of the motor

Q = √(S^2 - P^2) = √(34,641^2 - 29,445^2) ≈ 17,473 VAR

The power losses (P_{Loss}) in the motor

$P_{Loss} = P / \eta = 29,445 / 0.92 \approx 32,016$ W

Answers:

S ≈ 34,641 VA, P ≈ 29,445 W, Q ≈ 17,473 VAR&P_{Loss} ≈ 32,016 W

PRACTICAL

1. By varying the speed of the induction motor plot its Torque-Speed Characteristics. Also, study the performance and provide insights into starting, full load, and maximum torques.

2. Create a phasor diagram for an induction motor by representing the stator and rotor currents and voltages as phasors. Observe how these phasors vary with the motor's operating conditions (start-up, full load, etc.). This practical helps understand the electrical aspects of the motor and how they relate to its mechanical behaviour.

3. Employ a variable frequency drive (VFD) to control the speed and direction of an induction motor. Run the motor in all four quadrants (forward motoring, forward regenerative braking, reverse motoring, and reverse regenerative braking). This experiment showcases the motor's capability to operate in different directions and helps understand the concept of four-quadrant operation.

4. Set up an induction motor with a load and measure the input power, output power, and losses. Create a power flow diagram to visualize how electrical power is converted into mechanical power and any losses incurred in the process. This practical provides insights into motor efficiency and the distribution of power within the system.

KNOW MORE

1. Power Factor Explained - The basics what is power factor pf
   https://youtu.be/Tv_7XWf96gg?si=D1mEDo62LIRNb6fp
2. Types of Motors Used in Industrial Designs
   https://youtu.be/EW2uQnRmUIQ?si=sGX-_2xLvDWYgdly

REFERENCES AND SUGGESTED READINGS

1. P.S. Bimbhra, Electric Machines, Khanna Book Publishing Co., New Delhi (ISBN: 978- 93-6173- 294)
2. Mittle, V.N. and Mittle, Arvind., Basic Electrical Engineering, McGraw Hill Education New Delhi, ISBN :9780070593572
3. Kothari, D. P. and Nagrath, I. J., Electrical Machines, McGraw Hill Education. New Delhi, ISBN:9780070699670
4. Bhattacharya, S. K., Electrical Machines, McGraw Hill Education, New Delhi, ISBN:9789332902855
5. Theraja, B.L., Electrical Technology Vol-II (AC and DC machines), S.Chand and Co. Ltd., New Delhi, ISBN : 9788121924375
6. Sen, S. K., Special Purpose Electrical Machines, Khanna Publishers, New Delhi, ISBN: 9788174091529

7. Janardanan E. G, Special Electrical Machines, Prentice Hall India, New Delhi ISBN: 9788120348806
8. Hughes E., Electrical Technology, ELBS
9. Cotton H., Electrical Technology, ELBS

Dynamic QR Code for Further Reading

3 THREE-PHASE INDUCTION MOTOR- Starting and Speed Control

UNIT SPECIFICS

Through this module we have discussed the following aspects:

- *Importance of starters and familiarizing various starters of Induction motor;*
- *Various conventional and static speed control methods of Induction Motors;*
- *Selection of motor for various applications;*
- *Importance of motor maintenance;*
- *Various ways for maintenance:*
- *Problems.*

This Module delves into the industrial applications of various topics to bring curiosity and foster creativity. It includes different types of assessments, such as multiple-choice, short answer, and long answer questions, categorized according to Bloom's taxonomy. Additionally, the unit provides numerical problems for practice, a list of references, and suggested readings. Notably, the unit incorporates QR codes throughout its sections, allowing readers to scan them for supplementary knowledge on specific topics of interest.

Following each practical aspect, there is a "Know More" section carefully crafted to provide users with additional valuable information. This section encompasses various elements, including introductory activities, intriguing facts, historical context related to subject development, noteworthy observations and discoveries, timelines tracing the topic's evolution to the present day, real-life and industrial applications, case studies involving environmental, sustainability, social, and ethical considerations, and intriguing and curiosity-inducing topics.

RATIONALE

Induction motors are the work horse in all the industries. It has extensive use in both industrial and residential applications. About 50% of the power consumption is due to the induction motor loads. This chapter introduces the students to importance of starting, speed control and maintenance of the motor. the rationale for starting, speed control, and maintenance of three-phase induction motors is to ensure safe and efficient operation, protect the electrical system from excessive current, meet varying load demands, reduce energy consumption, extend motor life, and prevent costly downtime. These practices are essential across a wide range of applications, from industrial manufacturing to residential appliances, to ensure the seamless functioning of electric motors.

PRE-REQUISITES

Machines I: Sem III

Physics: Electromagnetism (Class X)

UNIT OUTCOMES

List of outcomes of this unit is as follows:

U3-O1: Understand need of starters, various starting methods

U3-O2: Able to understand various speed control methods

U3-O3: How to select motor for various applications

U3-O4: Understand importance of maintenance

U3-O5: Able to get to know various ways to execute maintenance

Unit-3 Outcomes	EXPECTEDMAPPINGWITHCOURSEOUTCOMES (1-WeakCorrelation;2-Mediumcorrelation;3-StrongCorrelation)				
	CO-1	CO-2	CO-3	CO-4	CO-5
U3-O1	-	-	3	-	-
U3-O2	-	-	3	-	-
U3-O3	-	-	3	-	-
U3-O4	-	-	3	-	-
U3-O5	-	-	3	-	-

3.1 INTRODUCTION

Three-phase induction motors, often referred to as asynchronous motors, are among the most ubiquitous and indispensable components in the realm of electrical machinery. Their significance lies in their exceptional reliability, efficiency, and adaptability to a myriad of industrial, commercial, and residential applications. Operating on the fundamental principles of electromagnetic induction, these motors exhibit a straightforward yet robust design, consisting of a stator with three windings supplied with three-phase alternating current (AC) power and a rotor, typically of the squirrel-cage type, which remains unconnected to an external power source. What distinguishes these motors is their inherent ability to self-start, without the need for brushes or commutators, making them virtually maintenance-free. This self-starting characteristic, combined with their ability to generate substantial torque, renders three-phase induction motors the ideal choice for applications requiring the conversion of electrical energy into mechanical work, such as driving pumps, fans, conveyors, compressors, and various other industrial machinery.

Their versatility in adapting to varying loads and requirements positions three-phase induction motors as the driving force behind the productivity of countless industries. Their role extends from propelling heavy machinery in manufacturing plants to ensuring efficient circulation in HVAC systems, and from powering essential appliances in households to maintaining water distribution in municipal utilities. Beyond their reliability and efficiency, three-phase induction motors offer another crucial advantage: their ease of integration with advanced control systems and speed-regulating devices, allowing for precise management of motor speed and performance in applications that demand accuracy and efficiency. As such, these motors serve as the cornerstone of modern automation and industrial processes, contributing to enhanced productivity, energy conservation, and the seamless operation of vital systems in our daily lives.

3.2 STARTERS

The motor starters are the device used for starting of motor. Its primary purpose is to limit the high inrush (starting) current while switch on the motor. In general, the starting current is 4 to 10 times depends on the motor rating and can cause voltage drops in mains, stress on the electrical system, and mechanical stress on the motor itself.

A significant inrush current occurs at the time of starting in an induction motor due to the motor's inherent characteristics and the physics of its operation. Several factors contribute to this phenomenon:

a) Initial Magnetization of the Core: The magnetic core of the motor is initially unmagnetized. The absence of magnetic flux results in very low impedance in the motor windings. Consequently, when the motor is switched on the magnetic field begins to build up in the core due to this initial current drawn is exceptionally high, causing a momentary surge.

b) Lack of Counter Electromotive Force (Back EMF): During startup, the motor rotor is stationary. As a result, there is no rotation to induce a counter electromotive force (back EMF) in the motor windings.
c) High Torque Requirement: To overcome inertia and initiate motion, the motor needs a high torque. The mechanical Torque in an induction motor is directly proportional to the current drawn from the supply, so a high inrush current is required to produce the necessary starting torque.
d) Impedance Mismatch: The impedance (resistance and reactance) of the motor windings and the supply system may not match perfectly at startup. This mismatch can result in a momentary spike in current as the motor and power supply adjust to each other's characteristics.
e) Voltage Drop: Voltage drop in the power supply system can worsen the inrush current issue. If the supply voltage is not perfectly constant, the motor may draw even more current to compensate for any voltage sag during startup.

It's important to note that while the inrush current is high during startup, it quickly decreases as the motor begins to rotate. As the rotor accelerates, it generates a back EMF that opposes the supply voltage, reducing the current to its normal operating value. In most cases, the inrush current lasts for only a fraction of a second.

To mitigate the effects of inrush current and its potential impact on the electrical system, various motor starters, such as star-delta starters, auto-transformer starters, and electronic motor starters, are employed. These devices provide controlled and gradual voltage ramp-up to the motor, reducing the initial current surge and minimizing voltage disturbances in the electrical system. There are several types of motor starters commonly used with three-phase induction motors:

1. Direct Online Starter (DOL)
2. Stator Resistance starter
3. Autotransformer Starter
4. Star Delta Starter
5. Soft Starter
6. Rotor Resistance or Slip Ring Motor Starter

Direct On-Line (D.O.L.) Starting

Fig.3.1 gives the circuit configuration for starting of 3Ø induction motor using DOL starter. The current drawn by the motor while starting about 4 to 7 times of the rated current even under no load condition. The starting torque also required will be 1.5 to 2.5 times the rated torque. The parts of a DOL starter is main contactor, protective devices, overload relay, start and stop push buttons which is used for motor starting operations. Pressing the start button energizes the contactor, which simultaneously connects the supply to three phase windings in stator. When the "STOP" button is pressed the main contact is released and supply to the motor is cut off.

DOL starters are suitable for motors with ratings below 5 kW to prevent excessive line voltage drop caused by the high starting current.

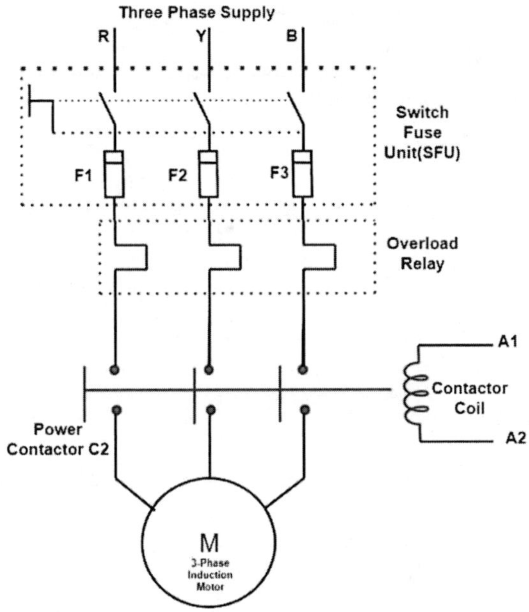

Fig.3.1: Schematic representation of D.O.L starting

Starting of Squirrel Cage Motors

To limit the inrush current in the squirrel cage induction motor the supply voltage to stator can be reduced, and the following are the different methods of it.

- Stator resistance
- Star-delta switches
- Autotransformer

1. Stator Resistance Starting

In this method, a variable external resistance is connected in series with each phase of the stator winding as shown in Fig.3.2 during starting. The connected external resistors reduces the voltage appeared to the motor as 70%. Thus, a reduced supply voltage applied across the motor terminals limiting the starting current. As the motor speed accelerated and reaches 80% of the rated speed gradually cut down the resistance fully so that the motor attains rated speed as rated voltage is applied to the motor. When the initial voltage is decreased by 50%, as per Ohm's law (V=IZ) dictates that the initial current will also reduce to the ratio. The relation between mechanical torque and input voltage of three-phase induction motor is the starting torque is proportional to the square of the applied voltage. Therefore, if the applied voltage is reduced to half of its rated value, the starting torque will only be ¼ th of its normal voltage value. This approach is commonly employed to achieve a smooth starting for small induction motors.

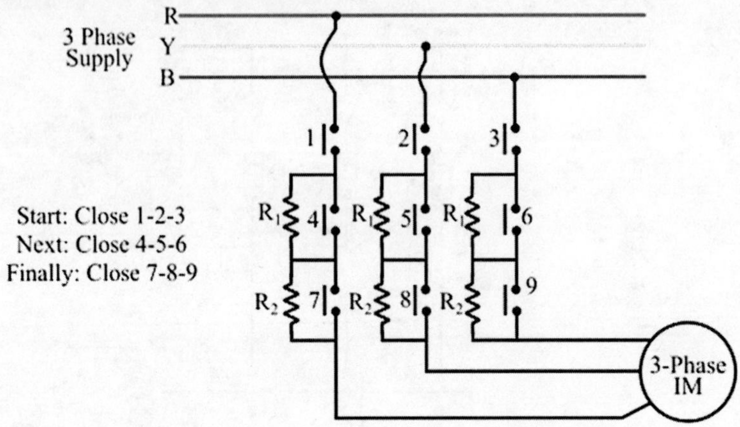

Fig.3.2: Schematic representation of stator resistance starting

Drawbacks:

- The starting torque is reduced and it increases acceleration time due to reduced voltage starting
- Power loss in the starting resistances.

2. Star-Delta Starting

Star-Delta Starter, also known as a wye-delta starter, is used for motors rated up to 25 HP.

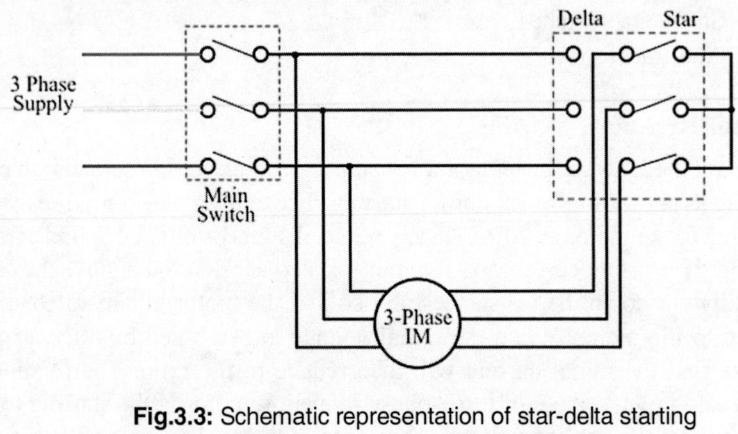

Fig.3.3: Schematic representation of star-delta starting

It starts the motor in a "star" (wye) configuration as shown in Fig.3.3, which reduces the voltage applied to the motor. After starting the motor, the speed increases and reaching certain value or time period the two-way switch moved to the "delta configuration, providing full voltage to the motor. Star-delta starters reduce the inrush current and mechanical stress on the motor When the stator windings are connected in start the voltage appeared across to the phase windings will be reduced by a factor $1/\sqrt{3}$ and the starting torque also reduced to 1/3 times compare to delta connected winding.

Disadvantage:

At the time of starting in the star mode there is a large reduction in stating torque.

3. Autotransformer Starting

Auto-transformers are compatible for both Y/Δ connected squirrel cage motors. They function as 3Ø step-down transformers, featuring various taps that enable motor starting at different voltage levels, such as 50%, 65%, or 80% of the rated voltage. When employing an auto-transformer as a starter, the current drawn by the motor while starting is lower than the rated current of the motor. The reduction in starting current is proportional to transformation ratio and output voltage of the auto transformer. For instance, when a motor starts on a 65% tap, it receives 65% of the line voltage and draws 65% of the line voltage's starting current value. However, the line current will be 42% (65% of 65%) of the line voltage's starting value, reflecting the difference due to the transformer's action. The internal connections of an autotransformer starter are illustrated in Figure 3.4.

During motor starting, the control switch is set to to the "start" position, in this case the auto transformer delivers the reduced voltage to motor and speed increases gradually.

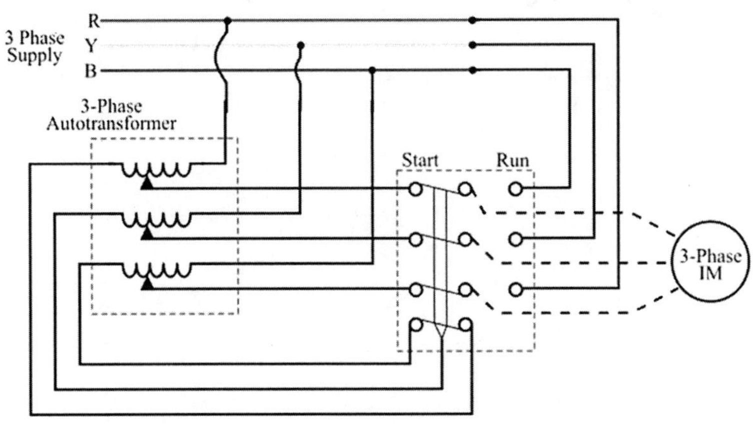

Fig.3.4: Schematic representation of autotransformer starting.

As the motor speed reaches 80% of its rated speed, then auto-transformer is automatically disconnected from the circuit when the switch is switched to the "run" position. The type of switch used for this transition can be either air-break (for small motors) or oil-immersed (larger motors).

Additionally, auto-transformer starters often incorporate safeguards against voltage drops and overloads, including time-delay circuits to ensure the motor operates efficiently and safely. The autotransformer starting method of induction motors has many advantages such as this method reduces the losses in starting, low starting current etc. This method is expensive and used for the motors rated above 25 HP.

Electronic Soft Starter

Electronic starters, or soft starters, are typically used with squirrel cage induction motors rather than slip ring induction motors. Electronic soft starters use power semiconductor switches to control the voltage supplied to the motor during startup. They provide a gradual voltage ramp-up, resulting in a smooth and controlled acceleration. Electronic soft starters are often used for motors of various sizes and are highly effective in reducing inrush current and mechanical stress. In a conventional motor starter, providing the motor with full voltage at the outset produces the highest starting torque, which can pose a mechanical risk to the motor.
Electrical soft starters controls and varies the supply voltage or current. Typically, voltage control is achieved through the use of anti-parallel connected Silicon Controlled Rectifiers (SCR) as depicted in Figure 3.5.

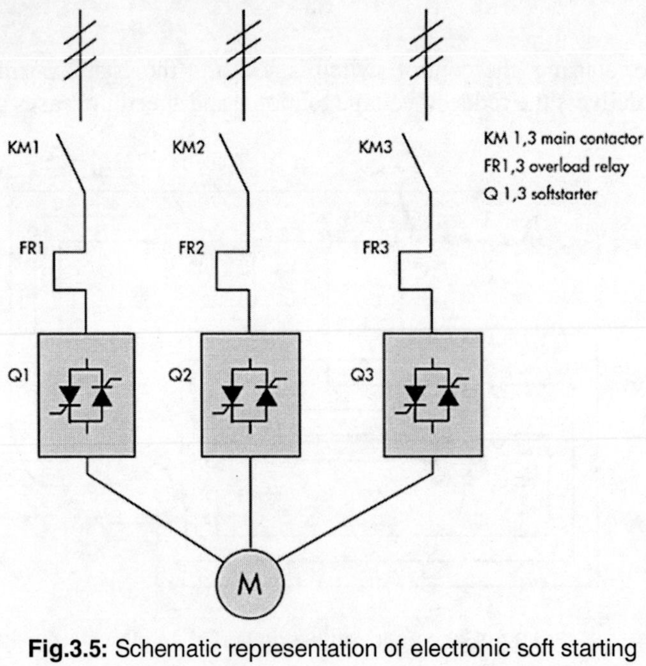

Fig.3.5: Schematic representation of electronic soft starting

Advantages of Soft Starter

- They provide very gradual increase of voltage and hence a smooth rise in the speed.
- By adjusting the firing angle, gradually or rapidly it controls the motor speed during starting and shutdown.
- They limit the inrush current and thus, preventing the power surges which avoids the overheating of motor.
- Since the losses in electronic starters are very less compared to resistive starters. So, the efficiency of the motor increased.
- The soft starters are smaller in size and compact that takes up very small space in mounting in motor control panels.
- Low Cost

Disadvantages of Soft Motor Starter

The electronic starters reduces the voltage applied to the motor but starting torque is directly proportional to the supply voltage the induction motor. So the s**oft starters are applicable for low or medium rated motors.**

Applications of Soft Starter

Industrial Fans, Conveyer belts, Motors using belt & pulleys

The choice of starter depends on factors such as:

- The motor's size,
- The load it drives,
- The available power supply, and
- The desired level of control during startup.

Slip Ring Induction Motor Starting or Rotor Resistance Starting

Slip-ring motors are started by applying the rated voltage to the stator, as additional start connected resistance are connected to the rotor circuit through slip-rings as shown in Figure 3.5. The external resistance added to the rotor side reduces the starting current in both the rotor and the stator, consequently enhancing power factor and increasing the starting torque.

This allows slip-ring motors can be started at full load condition. It's important to note that the external resistance is solely intended for starting purposes and is gradually removed as the motor gains speed. This method involves a star-connected variable resistance in the rotor circuit via slip-rings, as depicted in Figure 3.6.

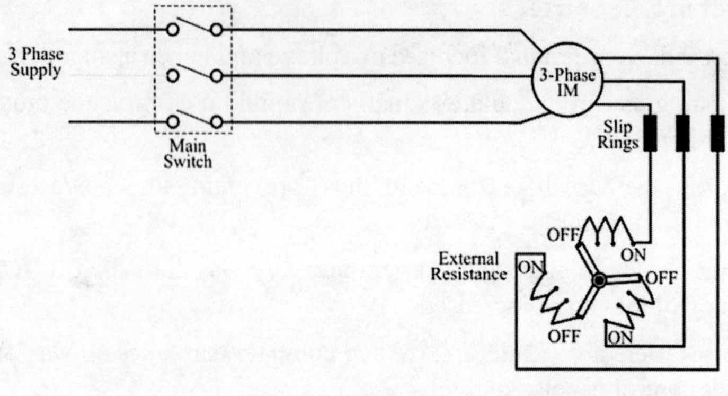

Fig.3.6: Schematic representation of rotor resistance starting.

3.3 SPEED CONTROL METHODS

Various industrial and commercial applications require precise control of motor speed and there are some common speed control methods, including stator voltage control, pole changing, rotor resistance control, and VVVF (Variable Voltage Variable Frequency) control. To get better clarity on speed control methods, let us see the basic formulas of speed and torque of three phase induction motor.

Synchronous speed (N_s) of the motor is given by

$$N_s = \frac{120f}{P}$$

where,

f - frequency of supply voltage and

P - total number of poles in stator

The rotor speed is given by:

$$N = N_s(1 - s)$$

where,

N - rotor speed in rpm.

N_s - synchronous speed in rpm.

s - is the slip.

The torque developed by a 3Ø induction motor is given by:

$$T = \frac{3}{2\pi N_s} \frac{sE_2{}^2R_2}{(R_2{}^2 + (sX_2)^2)}$$

When the rotor is at standstill slip, s =1.

So, the Torques equation is given by,

$$T = \frac{3}{2\pi N_s} \frac{{E_2}^2 R_2}{({R_2}^2 + (X_2)^2)}$$

where,

E_2 - the rotor emf.

N_s - the synchronous speed.

R_2 - the rotor resistance.

X_2 - the rotor inductive reactance.

3.3.1Speed Control from Stator Side

The Speed control of Induction Motor can be implemented at both stator and rotor side. The stator side speed control techniques are,

- V / f control or frequency control.
- Controlling supply voltage.
- Changing the number of stator poles.
- Adding rheostat in the stator circuit.

a) V / f Control or Frequency Control

The synchronous speed of the motor is given by

$$N_s = \frac{120f}{P}$$

The emf induced in the stator of the three-phase induction motor is given by:

$$E \; or \; V = 4.44 \, \emptyset K \, T_{ph} \, f$$

$$\varphi = \frac{V}{4.44KT_{ph}f}$$

where,

K - is the winding constant,

T_{ph} - is the number of turns per phase,

f - is the supply frequency.

Now if the frequency of supply voltage is reduced then the flux increases causing saturation of rotor and stator cores which will further cause increase in no load current of the motor. Hence, it is important to maintain flux, φ constant which is possible only by varying the voltage. To achieve the speed control in a three-phase induction motor using the V/f method, it's necessary to provide the motor with variable voltage and variable frequency. This can be conveniently achieved by employing a converter and inverter combination.

b) Controlling Supply Voltage

In this method, the voltage supplied to the stator windings of the motor is varied to control the speed. This method is simple and cost-effective but may not provide very precise speed control, especially at lower speeds.

The torque of a three-phase induction motor is given by:

$$T \propto \frac{sE_2{}^2 R_2}{(R_2{}^2 + (sX_2)^2)}$$

In low slip region $(sX_2)^2$ is negligibly small as compared to R_2. So, torque becomes:

$$T \propto \frac{sE_2{}^2}{R_2}$$

Since rotor resistance, R_2 is constant the equation of torque becomes:

$$T \propto sE_2{}^2$$

As the rotor induced emf $E_2 \propto V$, $T \propto sV^2$.

However, with decrease in the supply voltage, torque decreases. But for supplying a constant load, the slip should be increased which can be achieved by reducing the speed. This method of speed control sees limited applications in practice because it demands a significant reduction in voltage to achieve even slight speed changes. Consequently, this results in an elevated motor current, leading to overheating of the induction motor.

c) Pole Changing:

Some three-phase induction motors are designed with multiple sets of stator windings, each with a different number of poles. By switching between these windings, the motor's synchronous speed and, consequently, its operating speed can be changed. This method is well-suited for squirrel cage induction motors because these motors inherently develop rotor poles that match the stator winding's poles. By changing the total number of poles in stator, the Synchronous speed (N_s) can be varied. For instance, a stator may be wound with two 3-phase windings: one for 4 poles and the other for 8 poles, with a supply frequency of 50 Hz.

- synchronous speed when 4 pole winding is connected, Ns = 120*50/4 = 1500 RPM
- synchronous speed when 6 pole winding is connected, Ns = 120*50/8 = 750 RPM

3.3.2 Speed Control from Rotor Side

The speed control of three phase induction motor from rotor side are further classified as:

- Adding external resistance on rotor side.

- Cascade control method.
- Injecting slip frequency emf into rotor side.

a) Rotor Resistance Control:

An external resistance is introducing in series to the rotor circuit of the motor. By varying the resistance, the rotor current and, hence, the motor speed can be controlled. Rotor resistance control is often used in wound rotor induction motors and provides good speed control at the cost of increased complexity and energy dissipation in the resistors. This is similar to armature control method in DC motor speed control.

b) Cascade Operation

In this method, two motors are mechanically coupled on a common shaft to ensure they operate at the same speed. One of these motors is supplied by a 3φ supply, while the other motor is driven by the induced electromotive force (emf) generated in the first motor through slip-rings. as shown in Fig.3.7.

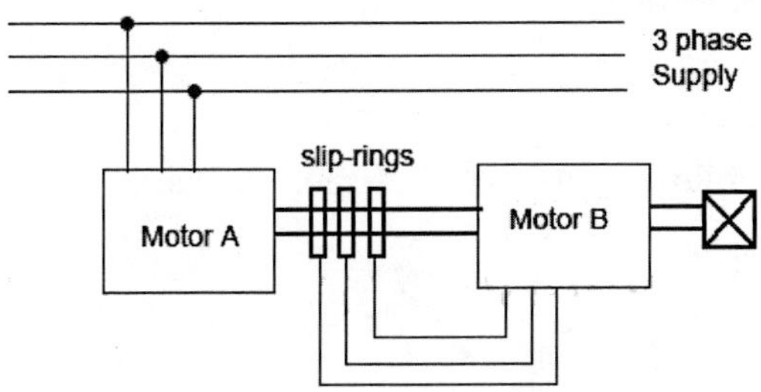

Fig.3.7: Speed control of slip ring induction motor via cascade operation

Motor A is called the main motor and motor B is called the auxiliary motor.

Let, N_{s1} = Supply frequency of motor A

N_{s2} = Supply frequency of motor B

P_1 = No. of stator poles of motor A

P_2 = No. of stator poles of motor B

N = Speed of the set and same for both motors

f = Frequency of the supply

Now, slip speed of motor A,

$$s_1 = \frac{N_{s1} - N}{N_{s1}}$$

Frequency of the rotor induced emf in rotor A

$$f_1 = s_1 f$$

Now, auxiliary y motor B is supplied with the rotor induce emf therefore,

$$Ns_2 = \frac{120\ f_1}{P_2} = \frac{120\ s_1 f}{P_2}$$

Now substitute the value of s_1

$$Ns_2 = \frac{120\ f\ (N_{s1} - N)}{P_2 N_{s1}}$$

At no load, speed of the auxiliary rotor is almost same as its synchronous speed.
i.e. $N = N_{s2}$.

From the above equations, it can be obtained that

$$N = \frac{120\ f}{(P_1 \pm P_2)}$$

where,

+ - Cumulative cascading

- = differential cascading

In this method the following speed ranges can be obtained

i. Motor A alone

$$NS_1 = \frac{120\ f}{P_1}$$

ii. Motor B alone

$$NS_2 = \frac{120\ f}{P_2}$$

iii. Cumulative Cascading

$$N = \frac{120\ f}{(P_1 + P_2)}$$

iv. Differential Cascading

$$N = \frac{120\ f}{(P_1 - P_2)}$$

c) By Injecting EMF In Rotor Circuit

In this technique speed of an induction motor is varied by external voltage applied into the rotor circuit. It's crucial that the injected voltage (electromotive force or emf) matches the same frequency as the slip frequency. However, there are no restrictions on the phase of the injected emf. When we inject emf that is in opposition to the phase of the rotor-induced emf, it increases the rotor resistance. Conversely, when we inject emf that is in phase with the rotor-induced emf, it reduces the rotor resistance. Thus, by altering the phase of the injected emf, speed control can be achieved. This method offers the significant advantage of achieving a wide range of speeds, both above and below the rated speed range. Various methods can be used to inject emf, including systems like the Kramer system and Scherbius system.

3.4 VARIABLE VOLTAGE VARIABLE FREQUENCY (VVVF) CONTROL

The VVVF Drive is a type of induction motor drive used for speed control in industries. This method is mainly used to control the speed or torque of three-phase induction motors. VVVF control, also known as VFD (Variable Frequency Drive) or VSD (Variable Speed Drive), is one of the most versatile and efficient methods for motor speed control. This method varies both the voltage (V) and frequency (f) supplied to the motor and thus it controls of motor speed and torque as required. They are widely used in various applications, including pumps, fans, conveyor systems, and HVAC systems.

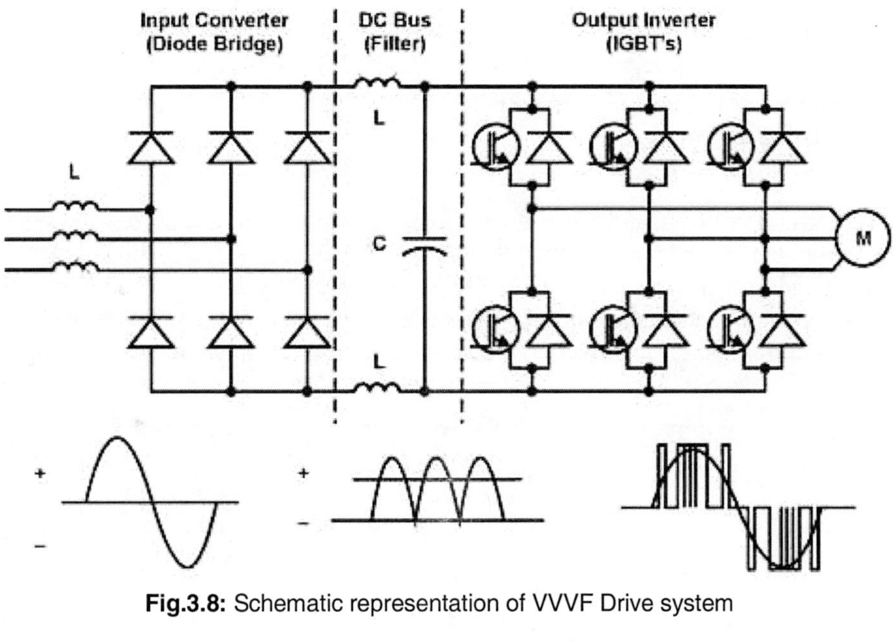

Fig.3.8: Schematic representation of VVVF Drive system

The VVVF Drive system has three main blocks as shown in Fig.3.8. They are:

1. Voltage Controlled Rectifier or AC-DC Converter
2. DC Link or Filter
3. Frequency Controlled Inverter

VFD's convert incoming AC power to DC and then use an inverter to generate the desired output voltage and frequency.

Voltage Controlled Rectifier or AC-DC Converter

The rectifier circuit converts the input three phase AC voltage into DC voltage. It's referred to as a "voltage-controlled rectifier" because, beyond converting AC to DC, it can also control or vary the rectified output voltage level. This converter or rectifier circuit is typically constructed using components like diodes, silicon-controlled rectifiers (SCRs), among others. By modulating the gate pulse sent to the SCRs, it's possible to adjust both voltage and current conduction.

Filter Circuit or DC Link

It contains either a parallel/series capacitor and inductor or a combination of both. The main purpose of this circuit is to smooth or filter the ripple in rectified DC output voltage from and supplies the constant DC link voltages to the inverter circuit.

Inverter

The inverter circuit plays an important role and it converts applied DC voltage into a variable three-phase AC supply. The output voltage and frequency are controlled by pulse width modulation techniques.

The variable voltage variable frequency output from the inverter is applied to induction motor to control the speed. Typically, this inverter circuit is consisting high-speed switching semiconductor devices such as MOSFETs or IGBTs. By selecting suitable inverter circuit and PWM control methods, control the voltage and frequency of the input AC supply. The controller and control logic for the inverter circuit is implemented using a microprocessor.

3.5 SELECTION OF INDUCTION MOTOR FOR VARIOUS APPLICATIONS

Selecting the right three-phase induction motor for different applications involves considering the torque-speed requirements of the load. Matching the motor characteristics to the load characteristics is crucial for efficient and reliable operation. Here's a general guide for selecting three-phase induction motors based on different load torque-speed requirements:

a) Constant Speed Applications:

In applications where the load requires a constant speed regardless of variations in the load torque, a standard squirrel-cage induction motor is suitable. Choose a motor with a nameplate speed that closely matches the desired operating speed.

b) Variable Torque Applications:

For applications with variable torque loads, such as centrifugal pumps and fans, consider motors with the following characteristics:

- Select a motor with a speed rating that closely matches the desired operating speed.
- Look for a motor with a horsepower rating appropriate for the load's maximum torque requirements.
- Ensure that the motor's service factor (usually mentioned on the nameplate) is sufficient to handle temporary overload conditions.

c) Constant Torque Applications:

In applications with constant torque loads, such as conveyor systems and extruders, focus on motors

- That can deliver consistent torque over a wide speed range.
- With a higher horsepower rating than required for the load's torque requirements at rated speed.
- Designed for inverter-duty or VVVF control, as these motors can handle varying speeds while providing constant torque.

d) High Starting Torque Applications:

Some loads, like crushers and compressors, require a high starting torque to overcome initial resistance. Select motors with a high starting torque. Motors designed for high torque applications typically have a lower slip, meaning they can handle higher loads during startup.

e) Variable Speed Applications:

For applications that require variable speed control, such as conveyors with changing load conditions, use Variable Frequency Drives (VFDs) in combination with standard induction motors. Choose a motor with a nominal speed rating and horsepower that meet the load's requirements under typical operating conditions. Use a VFD to adjust the motor's speed as needed to match the load's torque-speed curve.

f) Specialized Motors:

Some applications may require specialized motors, such as explosion-proof motors for hazardous environments or washdown-duty motors for wet environments. Ensure that the motor selected meets the specific requirements and safety standards of the application.

g) Efficiency Considerations:

Consider energy-efficient motors, such as those labelled as "NEMA Premium" or "IE3" for improved energy savings, especially in applications with constant or variable torque loads. Always refer to the load's torque-speed requirements, and consult motor manufacturers' catalogues and technical specifications to choose the most appropriate three-phase induction motor for a specific application. Additionally, compliance with local regulations, safety standards, and environmental conditions should also be considered during the selection process.

3.6 MAINTENANCE OF THREE PHASE INDUCTION MOTOR

Regular and proactive maintenance practices can significantly extend the lifespan of three-phase induction motors and reduce the risk of unexpected failures. Neglecting maintenance can lead to motor failures, costly downtime, and reduced productivity. Additionally, adhering to manufacturer recommendations and industry standards is essential to maintain motor reliability and efficiency.

Here is some key maintenance practices for three-phase induction motors:

- Visual Inspection: Regularly inspect the motor for signs of wear, damage, or overheating. Look for loose or damaged connections, frayed wires, and any unusual noises or vibrations.
- Cleaning: Keep the motor clean and free from dust, dirt, and debris. A clean motor dissipates heat more efficiently and is less likely to overheat.
- Lubrication: If the motor has lubrication points (e.g., for bearings), follow the manufacturer's recommendations for lubrication intervals and use the appropriate lubricant.
- Alignment: Ensure that the motor and driven equipment are properly aligned. Misalignment can lead to excessive vibrations and premature bearing failure.
- Bearing Maintenance: Inspect and lubricate motor bearings regularly. Replace bearings when they show signs of wear, noise, or excessive play. Monitor bearing temperature using temperature sensors, if available, to detect overheating.
- Cooling System: Check the motor's cooling system, such as fans and cooling fins, for proper operation. Ensure that the motor is adequately cooled to prevent overheating.
- Electrical Connections: Inspect electrical connections and terminals for signs of corrosion, loose connections, or overheating. Tighten loose connections and replace damaged components.
- Insulation Resistance Testing: Periodically perform insulation resistance tests to check the condition of the motor windings. This helps detect insulation breakdown before it leads to motor failure.
- Vibration Analysis: Use vibration analysis tools to monitor motor vibrations. Unusual vibration patterns can indicate problems with the motor or its mounting.
- Regular Maintenance Schedule: Develop a maintenance schedule based on the motor's operating conditions and load. High-load motors may require more frequent maintenance.
- Environmental Considerations: Protect the motor from harsh environmental conditions, such as extreme temperatures, moisture, and corrosive substances. Use appropriate enclosures or covers when necessary.
- Balancing: For motors with rotating parts like fans or rotors, ensure that they are balanced to reduce vibrations and extend the life of the motor and connected equipment.

- Infrared Thermography: Periodically perform infrared thermography scans to identify hot spots or overheating issues in the motor, which can indicate problems with electrical connections or winding insulation.
- Keep Records: Maintain detailed records of maintenance activities, including inspection dates, repairs, and replacements. This documentation can help track the motor's history and plan future maintenance.
- Training and Education: Ensure that maintenance personnel are properly trained in motor maintenance procedures and safety practices.

UNIT SUMMARY

This comprehensive unit provides a holistic understanding of starters in three-phase induction motors, their control methods, selection criteria, and maintenance practices, enabling effective application and upkeep in diverse industrial settings.

EXERCISES

1. Given a 10 HP three-phase induction motor that drives a high-inertia load. Explain why using a soft starter would be a better choice than a DOL starter for this application. Calculate the approximate inrush current reduction achieved by using a soft starter compared to a DOL starter for this motor.

2. A manufacturing plant has a variety of induction motors of different sizes. Describe the specific criteria and considerations that should guide the choice of starter type for each motor based on its size and application. Create a table listing different motor sizes and their recommended starter types, explaining the rationale behind each choice.

3. In a material handling system, you need to control the speed of a conveyor belt driven by a three-phase induction motor. Discuss the advantages and disadvantages of using stator voltage control compared to VVVF control using a VFD for this application. Calculate the potential energy savings by using VVVF control compared to stator voltage control over a year if the conveyor operates 24/7.

4. You are tasked with selecting a motor for a water pump used in a remote agricultural location with unreliable power supply. Describe the motor characteristics, control method, and maintenance considerations for this specific application.

5. A company is expanding its production line and needs to choose motors for various machines. Explain the importance of load torque-speed requirements and how they affect motor selection for different machines in the production line.

6. A facility manager is implementing a preventive maintenance program for a set of induction motors. Describe the key steps and tasks that should be included in the maintenance checklist for these motors.

7. During a routine inspection, you discover an unusual vibration in a motor. Discuss the possible causes of this vibration and the steps you would take to diagnose and resolve the issue. Provide a list of common vibration analysis tools and their specific uses in diagnosing motor issues.

Multiple Choice Questions

1. Why are starters used in three-phase induction motors?
   a) To provide mechanical support
   b) To control motor speed
   c) To protect against overload
   d) To improve power factor

2. Which type of starter is commonly used for reducing the starting current of a motor?
   a) DOL starter
   b) Star-delta starter
   c) Auto-transformer starter
   d) Soft starter

3. What is the primary purpose of stator resistance starters?
   a) To reduce motor speed
   b) To protect the motor against high voltage
   c) To increase the motor's power factor
   d) To limit the starting current

4. Which method allows for changing the number of poles in a three-phase induction motor?
   a) Stator voltage control
   b) Pole-changing method
   c) Rotor resistance control
   d) VVVF control

5. VVVF stands for:
   a) Very Variable Voltage Frequency
   b) Variable Voltage Variable Frequency
   c) Voltage Variation via Frequency

d) Variable Voltage Variable Flux

6. Which type of motor is typically suitable for applications requiring high starting torque?
   a) Synchronous motor
   b) Induction motor
   c) Stepper motor
   d) DC motor

7. In which application would you typically choose a motor with a variable frequency drive (VFD)?
   a) Pump
   b) Ceiling fan
   c) Conveyor belt
   d) Electric oven

8. What is the primary purpose of lubricating the motor bearings during maintenance?
   a) To increase motor speed
   b) To reduce motor noise
   c) To prevent bearing wear and extend motor life
   d) To improve motor efficiency

9. Which of the following is an essential maintenance task for three-phase induction motors?
   a) Changing the motor's voltage rating
   b) Periodically checking the motor's power factor
   c) Cleaning the motor's exterior casing
   d) Checking and tightening electrical connections

Answers of Multiple-Choice Questions

1. c) To protect against overload
2. b) Star-delta starter
3. d) To limit the starting current
4. b) Pole-changing method
5. b) Variable Voltage Variable Frequency
6. b) Induction motor

7. c) Conveyor belt
8. c) To prevent bearing wear and extend motor life
9. d) Checking and tightening electrical connections

Short and Long Answer Type Questions

Long Answer Questions with Answers

1. **Explain the need for starters in three-phase induction motors and discuss the different types of starters used.**
   Three-phase induction motors require starters for several reasons. One primary need is to limit the high inrush current that occurs when the motor is initially switched on. This inrush current can be several times higher than the motor's rated current and can cause voltage drops in the power supply system, leading to voltage fluctuations and potential damage to other connected equipment. Starters help mitigate this issue by gradually applying voltage to the motor, allowing it to start smoothly.
   There are several types of starters used in three-phase induction motors:
   - Direct-On-Line (DOL) Starter: This is the simplest type of starter and directly connects the motor to the power supply at full voltage. It is suitable for small motors with low starting torque requirements but can lead to a significant inrush current.
   - Star-Delta Starter: Star-delta starters are used for motors with higher starting torque requirements. They initially connect the motor in a star configuration (lower voltage) to
   - reduce the starting current and then switch to a delta configuration (full voltage) after a brief period.
   - Auto-Transformer Starter: Auto-transformer starters use an auto-transformer to reduce the starting voltage temporarily and then increase it gradually during startup. This method reduces inrush current and provides smoother acceleration.
   - Rotor Resistance Starter: Rotor resistance starters are used in slip-ring induction motors. They insert external resistance into the rotor circuit during startup to control starting current and torque.
   - Soft Starter: Soft starters use solid-state electronics to control the voltage applied to the motor, gradually ramping it up during startup. This results in a smooth and controlled acceleration, minimizing inrush current.

   Each type of starter is chosen based on the motor's characteristics, starting torque requirements, and the impact on the power supply system.

2. **Describe various methods for controlling the speed of three-phase induction motors and discuss their advantages and disadvantages.**
   Three-phase induction motor speed control is essential in various industrial applications. Several methods are employed for this purpose, each with its advantages and disadvantages:
   - Stator Voltage Control: By varying the voltage supplied to the stator windings, the motor's speed can be controlled. Reducing voltage decreases speed, and increasing voltage increases speed. However, this method may lead to reduced torque at lower speeds and can affect motor efficiency.
   - Pole Changing Method: This method involves changing the number of poles in the motor

by switching between multiple winding configurations. It allows for discrete speed changes but is limited by the available winding configurations and may lead to increased complexity and cost.

- Rotor Resistance Control: By inserting external resistance into the rotor circuit, the rotor current and torque can be controlled, affecting speed. This method is effective for slip-ring motors but results in energy losses in the resistors.
- Variable Voltage Variable Frequency (VVVF) Control: VVVF control is achieved using variable frequency drives (VFDs). By independently controlling voltage and frequency, VFDs provide precise speed control while maintaining torque across a wide range of speeds. This method is highly efficient but may require additional investment in VFD equipment.

The choice of speed control method depends on the application's specific requirements, including the desired speed range, torque, and efficiency goals.

3. **Discuss the factors to consider when selecting a three-phase induction motor for a specific application, taking into account load torque-speed requirements.**

   Selecting the right three-phase induction motor for a given application involves considering several factors, primarily related to the load's torque-speed requirements:

   - Load Torque Profile: Understanding the load's torque-speed profile is crucial. Some applications require high starting torque, while others need constant speed or variable speed control. This profile helps determine the appropriate motor type and control method.
   - Operating Environment: Consider the operating environment, including temperature, humidity, and the presence of dust or corrosive substances. Choose a motor that can withstand these conditions.
   - Load Inertia: The load's inertia affects the motor's ability to accelerate and decelerate. High-inertia loads may require motors with higher starting torque.
   - Speed Range: Determine the required speed range for the application. Some motors may have limitations on their speed control capabilities.
   - Power Supply: Ensure that the available power supply voltage and frequency match the motor's requirements.
   - Efficiency Requirements: Consider energy efficiency goals. High-efficiency motors may be preferred for applications where energy consumption is a concern.
   - Maintenance Requirements: Evaluate the maintenance requirements of the motor. Motors with sealed bearings and low maintenance needs are preferred for some applications.
   - Budget Constraints: Factor in budget limitations and cost considerations when selecting a motor and control system.

4. **Explain the key maintenance practices for ensuring the reliable operation of three-phase induction motors.**

   Maintenance is crucial for prolonging the life and ensuring the reliable operation of three phase induction motors. Key maintenance practices include:

   - Regular Inspection: Periodically inspect the motor for signs of wear, damage, or loose connections. This includes checking the housing, electrical connections, and cooling fans.
   - Lubrication: Lubricate motor bearings as recommended by the manufacturer. Proper

lubrication reduces friction, heat, and wear in the bearings.

- Cleaning: Keep the motor clean from dust, dirt, and debris. Clean the exterior and ensure proper ventilation to prevent overheating.
- Alignment and Balancing: Check and correct motor alignment to prevent vibration issues. Imbalanced loads can lead to premature wear.
- Temperature Monitoring: Use temperature sensors to monitor motor temperature. Abnormal temperature increases may indicate problems with the motor or its load.
- Electrical Testing: Periodically test the motor's insulation resistance, winding resistance, and electrical connections to detect any deterioration.
- Vibration Analysis: Employ vibration analysis tools to detect mechanical problems, such as misalignment or imbalance, which can lead to motor failure.
- Bearing Replacement: Replace motor bearings when they show signs of wear or damage to prevent motor breakdown.
- Regular Maintenance Records: Maintain records of all maintenance activities, including dates, repairs, and replacements. This helps in scheduling preventive maintenance.
- Safety: Always follow safety procedures and lockout/tagout protocols when performing maintenance to prevent accidents.

Long answer Questions:

1. Discuss the advantages and disadvantages of using an auto-transformer starter compared to a star-delta starter in three-phase induction motors. Provide practical examples where each type would be preferred.
2. Explain the principle of Variable Voltage Variable Frequency (VVVF) control in three-phase induction motors using a Variable Frequency Drive (VFD). Discuss the benefits of VVVF control in terms of motor performance and energy efficiency.
3. You are tasked with selecting a three-phase induction motor for a high-inertia conveyor system in an industrial plant. Describe the key factors you would consider in motor selection, including motor type, size, and control method, and justify your choices based on the specific requirements of the application.
4. Imagine you are the maintenance supervisor in a manufacturing facility that relies heavily on three-phase induction motors. Discuss the importance of creating a comprehensive maintenance schedule for these motors. Outline the steps involved in developing such a schedule and the benefits it brings to the facility's operations.
5. Explain the concept of predictive maintenance for three-phase induction motors. Describe the techniques and tools available for predictive maintenance, such as vibration analysis, thermal imaging, and online monitoring systems. Highlight the advantages of adopting predictive maintenance over traditional preventive maintenance strategies.

Short answer Questions:

1. What is the primary purpose of using a starter in a three-phase induction motor?
2. Name three common types of motor starters used for controlling the starting current of induction motors.
3. Explain the difference between a Direct-On-Line (DOL) starter and a Soft Starter.
4. What does VVVF stand for in the context of motor control?

5. Briefly describe the pole-changing method for controlling the speed of induction motors.
6. How does stator voltage control affect the speed of a three-phase induction motor?
7. What factors should be considered when selecting a motor for an application with high starting torque requirements?
8. Name one application where a variable frequency drive (VFD) is commonly used for motor speed control.
9. Why is it important to match the motor's power supply requirements to the available power source?
10. What is the purpose of lubricating the bearings in a three-phase induction motor?
11. Give an example of a maintenance task that helps prevent electrical faults in induction motors.
12. How can vibration analysis be used for motor maintenance, and what does it help detect?

Numerical Problems

1. Consider a 3-phase, 415V, 50Hz induction motor with the following specifications: Power rating: 7.5 kW; Full-load current: 15 A; Starting method: Direct-On-Line (DOL) starter. Calculate the following: a. Starting Current (Ist); b. Size of Overload Relay

Solution:

a. Starting Current (I_{st}):

In a DOL starter, the starting current is typically 6-8 times the full-load current.
Let's assume it's 7 times the full-load current:
Ist = 7 * 15 A = 105 A

b. Size of Overload Relay:

To protect the motor from overheating, we need to select an overload relay.
Typically, the relay setting is 125% of the full-load current. So,
Overload relay setting = 1.25 * 15 A = 18.75 A ≈20A

Answers:

Starting Current (Ist) = 105 A
Size of Overload Relay≈20A

2. Consider a 3-phase, 380V, 50Hz induction motor with these specifications: Power rating: 15 kW; Full-load current: 30 A; Starting method: Star-Delta (Y-Δ) starter. Calculate the following: a. Starting Current (Ist); b. Size of Overload Relay

Solution:

a. Starting Current (I_{st}):

In a Y-Δ starter, the starting current is reduced to 1/3 of the DOL starting current. Therefore,
I_{st} = (1/3) * (30 A) = 10 A

b. Size of Overload Relay:

As before, we need to protect the motor with an overload relay. Using the same 125% setting, we get:
Overload relay setting = 1.25 * 30 A = 37.5 A ≈ 40A

Answers:

Starting Current (Ist) = 10 A
Size of Overload Relay = 40A

3. Suppose you have a 3-phase, 415V, 50Hz induction motor with the following specifications: Power rating: 15 kW; Synchronous speed (Ns): 1500 RPM; Full-load speed: 1440 RPM; Rated current: 30 A. The speed of the motor is regulated to 1200 RPM using VFD. Calculate the following: Frequency Setting for VFD and Motor Speed at 50 Hz.

Solution:

a. Frequency Setting for VFD:

Synchronous speed:

$N_s = (120 * f) / P$

Where N_s is the synchronous speed (1500 RPM), f is the frequency, and P is the number of poles.

Rearranging the formula to solve for frequency (f):

f = (Ns * P) / 120

For this motor with 1500 RPM synchronous speed and 4 poles:

f = (1500 * 4) / 120 = 50 Hz

So, you would set the VFD to output a frequency of 50 Hz to achieve 1200 RPM.

b. Motor Speed at 50 Hz:

Now, calculate the motor's speed at the selected frequency of 50 Hz:

Motor Speed = (f / 50 Hz) * Ns

Motor Speed = (50 / 50) * 1500 RPM = 1500 RPM

The motor will run at 1500 RPM when the VFD is set to 50 Hz.

Answers:

Frequency Setting for VFD = 50 Hz

Motor Speed at 50 Hz = 1500 RPM

4. Suppose you have a 3-phase, 415V, 50Hz induction motor with the following specifications: Power rating: 7.5 kW; Synchronous speed (Ns): 1500 RPM; Full-load speed: 1440 RPM Rated current: 15 A; Motor has two stator windings: 2 poles and 4 poles Calculate the speed when operating in 4-pole mode.

Solution:

a. Speed in 4-Pole Mode:

In 4-pole mode, the motor will have half the synchronous speed (Ns/2). So:

Speed = Ns / 2 = 1500 RPM / 2 = 750 RPM

When operating in 4-pole mode, the motor will run at 750 RPM.

Answers:

Speed in 4-Pole Mode = 750 RPM

PRACTICAL

1. Observe starting current, voltage, and the time it takes for the motor to reach full speed with varying loads and analyse the starting characteristics of the induction motor using a DOL starter.

2. Compare the starting characteristics by measuring starting current, voltage, and the time it takes for the motor to reach full speed when a star-delta starter is in star mode to that in delta mode.
3. Demonstrate the benefits of a soft starter in reducing starting current and mechanical stress during motor startup with DOL, autotransformer and start-delta starters.

KNOW MORE

1. Soft Starter for 3 Phase Induction Motors- full lecture!
   https://youtu.be/QCQFFUdYF94?si=jb3U7fL6zuKK9_Lk
2. What is a VFD? (Variable Frequency Drive)
   https://youtu.be/g7jFGOn6xfU?si=fZCbXuyxQ18lMqme
3. What is the Difference between VFD and Soft Starter?
   https://youtu.be/_ZztDN5XX5o?si=iEaTUjZJjwFUt2ec
4. Types of faults in three phase induction motor
   https://youtu.be/Zkbn62GALgI
5. Electric motor troubleshooting and maintenance techniques
   https://youtu.be/390nOrLHAaw?si=Y5OvOVU2Y4TIaGkX
6. Motor Maintenance & Troubleshooting [25KW Motor] | Motor Coil Winding Burned by Imbalance Rotor
   https://youtu.be/ntOc4h792UE?si=jUov0C0j4IO8jule

REFERENCES AND SUGGESTED READINGS

1. P.S. Bimbhra, Electric Machines, Khanna Book Publishing Co., New Delhi (ISBN: 978- 93-6173- 294)
2. Mittle, V.N. and Mittle, Arvind., Basic Electrical Engineering, McGraw Hill Education New Delhi, ISBN :9780070593572
3. Kothari, D. P. and Nagrath, I. J., Electrical Machines, McGraw Hill Education. New Delhi, ISBN:9780070699670
4. Bhattacharya, S. K., Electrical Machines, McGraw Hill Education, New Delhi, ISBN:9789332902855
5. Theraja, B.L., Electrical Technology Vol-II (AC and DC machines), S.Chand and Co. Ltd., New Delhi, ISBN : 9788121924375
6. Sen, S. K., Special Purpose Electrical Machines, Khanna Publishers, New Delhi, ISBN: 9788174091529
7. Janardanan E. G, Special Electrical Machines, Prentice Hall India, New Delhi ISBN: 9788120348806
8. Hughes E., Electrical Technology, ELBS
9. Cotton H., Electrical Technology, ELBS

Dynamic QR Code for Further Reading

4 SINGLE PHASE INDUCTION MOTORS

UNIT SPECIFICS

Through this module we have discussed the following aspects:

- *Familiarizing the working of single phase induction motor;*
- *Double field revolving theory;*
- *principle of making these motors self-start – Construction and working;*
- *Torque-speed characteristics;*
- *Motor selection for different applications as per the load torque-speed requirements;*
- *Maintenance of single phase induction motors;*

The photos of practical motor are shown to detail the various parts of the machine. The real time applications of the topics are discussed for inducing further curiosity and creativity.

Besides giving a large number of multiple-choice questions as well as questions of short and long answer types marked in two categories following lower and higher order of Bloom's taxonomy, assignments through a number of numerical problems, a list of references and suggested readings are given in the unit so that one can go through them for practice. It is important to note that for getting more information on various topics of interest some QR codes have been provided in different sections which can be scanned for relevant supportive knowledge.

After the related practical, based on the content, there is a "Know More" section. This section has been carefully designed so that the supplementary information provided in this part becomes beneficial for the users of the book. This section mainly highlights the initial activity, examples of some interesting facts, analogy, history of the development of the subject focusing the salient observations and finding, timelines starting from the development of the concerned topics up to the recent time, applications of the subject matter for our day-to-day life and industrial applications on variety of aspects, case study related to environmental, sustainability, social and ethical issues whichever applicable, and finally inquisitiveness and curiosity topics of the unit.

RATIONALE

Since home appliances are of single phase, three phase induction motor cannot be used. Industry applications with low power requirements are also in need of single phase only. About 50% of the power consumption is due to the induction motor loads. This chapter introduces the students to the construction and working principle of the motor. To understand the fundamentals in the principle of operation very important aspect is the starting methods of single phase induction motor and it's torque-speed characteristics.

PRE-REQUISITES

Machines I: Sem III

Physics: Electromagnetism (Class X)

UNIT OUTCOMES

List of outcomes of this unit is as follows:

U4-O1: Understand the construction of single-phase induction motor

U4-O2: Able to understand the working of different types of single-phase induction motor

U4-O3: Feel the production of double field revolving theory

U4-O4: Why self-starting is not possible in single phase induction motor

U4-O5: Able to get to know about the application of single-phase induction motor

Unit-4 Outcomes	EXPECTEDMAPPINGWITHCOURSEOUTCOMES (1-WeakCorrelation;2-Mediumcorrelation;3-StrongCorrelation)					
	CO-1	CO-2	CO-3	CO-4	CO-5	CO-6
U4-O1	3	-	-	-	-	-
U4-O2	3	-	-	-	-	-
U4-O3	-	3	-	-	-	-
U4-O4	-	3	-	-	-	-
U4-O5	-	3	-	-	-	-

4.1 INTRODUCTION

A single-phase induction motor is a type of electric motor that operates on a single alternating current (AC) supply. It is a widely used and versatile device, finding applications in various domestic and industrial settings due to its simplicity, reliability, and cost-effectiveness. Unlike three-phase motors that have three sets of windings and a rotating magnetic field, single-phase induction motors have only one set of windings, making their design and construction less complex.

The key principle behind the operation of a single-phase induction motor is the induction of a magnetic field in the stator (stationary part of the motor) when connected to an AC power source. The alternating current in the stator winding produces a magnetic field that alternates direction, creating a magnetic flux. This changing magnetic field induces a current in the rotor (rotating part of the motor) through electromagnetic induction, causing the rotor to turn and generate mechanical output.

Single-phase induction motors are commonly used in household appliances, such as fans, washing machines, and refrigerators, as well as in small-scale industrial applications. They are suitable for tasks that do not require extremely high power outputs, and their relatively simple construction makes them cost-effective and easy to maintain. Despite their straightforward design, engineers have developed various types and configurations of single-phase induction motors to optimize performance for specific applications.

In summary, single-phase induction motors play a crucial role in powering a wide range of devices, combining simplicity, efficiency, and affordability to meet the electrical motor needs of both residential and industrial environments.

4.2 SINGLE-PHASE INDUCTION MOTOR

For home and commercial applications, single-phase motors are more desirable than three-phase induction motors. Only single-phase electricity is available due to utility. Therefore, the three-phase induction motor cannot be used in this kind of application. And therefore, it is necessary to study about single phase induction motor. With the exception of the stator's single-phase winding, this motor's architecture is essentially identical to that of a three-phase induction motor. The rotor resembles the rotor of a squirrel cage induction motor with three phases. The end ring short circuits the rotor bars. An alternating flux is created when single-phase AC voltage is given to the stator coil.

The flux induces a voltage in the rotor during the positive half cycle, and the consequent current generates a torque. The rotor has a natural tendency to spin in one direction. The torque generated tends to rotate the rotor in the opposite direction during the negative half cycle. The rotor is incapable of rotating in any direction due to its inertia. The single phase induction motor cannot start on its own for this reason. Fig.4.1 depicts the single-phase induction motor's construction.

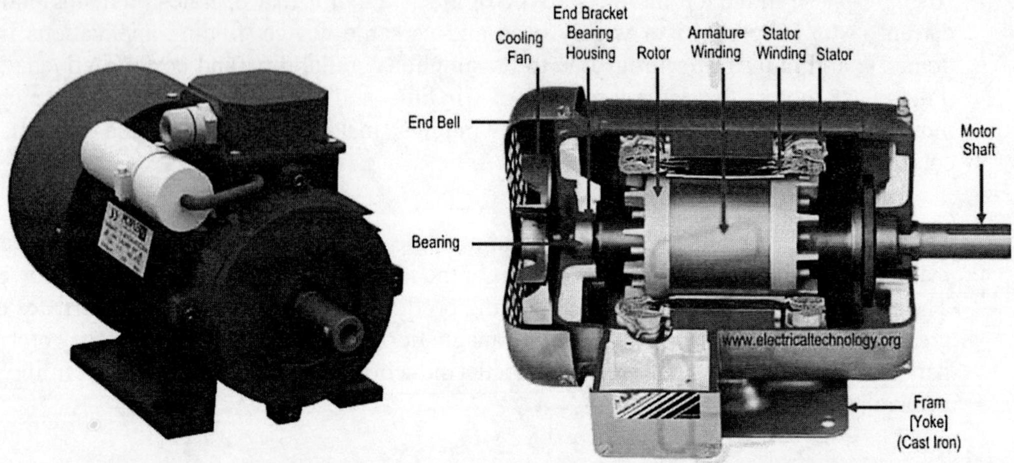

Fig.4.1: Construction of Single-Phase Induction Motor (Image Courtesy: Wikipedia)

Similar to a three-phase induction motor, a single-phase induction motor's two main parts are the Stator and Rotor. The only component that is different is the stator winding. The stator winding is a single phase, not three-phase. The stator core and the three-phase induction motor share the same core. A single-phase induction motor's stator has two windings, with the exception of shaded-pole induction motors. A primary winding and an auxiliary winding, respectively, make up these two windings. The stator core has laminated construction to reduce eddy current loss. The single-phase supply is delivered to the stator winding, also referred to as the main winding. The rotor of a single-phase induction motor is the same as that of a squirrel cage induction motor. Rotor winding is replaced with rotor bars, and the end rings short-circuit it. It consequently generates a complete path in the rotor circuit. The rotor bars are braced to the end rings to increase the mechanical strength of the motor. The rotor slots are in some way slanted to avoid magnetic coupling. It was also used to help a motor run smoothly and softly.

4.2.1 Working of Single-phase Induction Motor

The stator winding, often known as the main winding, receives a single-phase AC supply. Alternating current flowing through the stator coil produces magnetic flux. This flow is referred to as the "main flux". Imagine the rotor is moving while being encircled by a magnetic field produced by the stator windings. According to Faraday's law, when current first begins to flow in a circuit, it does so along a short path. This current is referred to as rotor current. The rotor current results in the production of rotor winding flux. This flux is known as rotor flux. There are two fluxes: the stator produces the primary flux, while the rotor generates the secondary flux. Interaction between the main flux and the rotor flux, resulting in a torque and rotation of the rotor. An alternating stator field is what it is. At the same synchronous speed, the stator field also moves. The frequency of the power source and the number of poles both affect the motor's synchronous speed. Using the double field revolving theory, the operation of a single phase induction motor may be understood. When the e.m.f. follows sine law, the flux follows cosine law given by $\Phi = \Phi_m \cos\omega t$

We know that

$$\cos \omega t = \frac{e^{jwt} + e^{-jwt}}{2}$$

$$\Phi = \Phi_m \frac{e^{jwt} + e^{-jwt}}{2}$$

$$\Phi = \Phi_m/2 \; e^{jwt} + \Phi_m/2 \; e^{-jwt}$$

Hence an alternating flux can be represented by two rotating fluxes, each equal to half the value and each rotating synchronously in opposite direction.

The resultant flux at different values of $\theta(\omega t)$ are explained using the following vector diagram.

$\theta = 0°$

When $\theta=0°$, resultant flux $= \frac{\Phi_m}{2} + \frac{\Phi_m}{2} = \Phi_m$ as shown in Fig.4.2 (a-i) when $\theta=30°$, $\Phi = 2x\frac{\Phi_m}{2}$ cos30 = 0.866Φm as shown in Fig.4.2 (a-ii). At $\theta=90°$, resultant flux is zero as shown in Fig.4.2 (a-iii). At $\theta= 180°$, Φ= Φm and $\theta=360°$, Φ=Φm as shown in Fig.4.2 (a-iv). The graph drawn with flux on y-axis and θ on x-axis is nothing but a Co-sinusoidal curve as shown in Fig.4.2 (b).

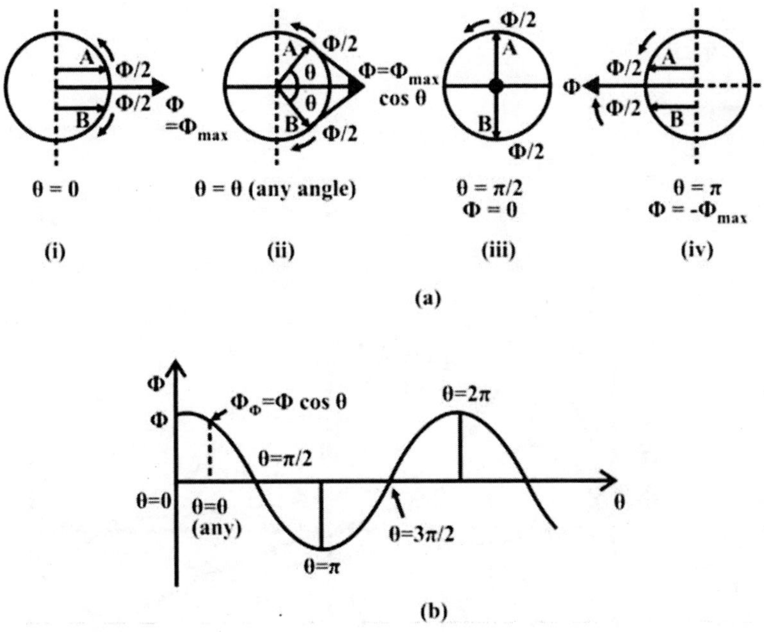

Fig.4.2: (a) Vector diagram, (b) Co-sinusoidal curve

If the actual speed of the rotor is N rpm, then the slip w.r.t. forward rotating flux is s_f.

$$s_f = \frac{N_S - N}{N_S}$$

The backward rotating field rotates in a direction opposite to the direction of the rotor. Hence the speed should be taken as –N.

$$s_b = \frac{N_S-(-N)}{N_S} = \frac{N_S+N}{N_S} = \frac{N_S+N-N_S+N_S}{N_S} = \frac{2N_S-(N_S-N)}{N_S}$$

$$s_b = 2 - s_f$$

Anticlockwise and clockwise fluxes revolving around the stator, cut the rotor conductors, induce a voltage in the rotor. This voltage circulates a rotor current as the rotor is short circuited. The forward field produces forward torque and the backward torque. The resultant torque is the difference of T_f and T_b, since T_f and T_b are in opposite directions.

We know that

Slip x rotor Input Power = Rotor Cupper loss

$$s_f \cdot T\omega_s = Ir^2 r_2$$

$$T = \frac{Ir^2 r_2}{s\omega_s}$$

Forward torque, $T_f = \frac{Ir^2 r_2}{s_f \omega_S}$

Net torque, $T = T_f - T_b$

$$T = = \frac{Ir^2 r_2}{\omega_S}\left(\frac{1}{s_f} - \frac{1}{s_b}\right)$$

$$T = K\, Ir^2 r_2 \left(\frac{1}{s_f} - \frac{1}{2-s_f}\right)$$

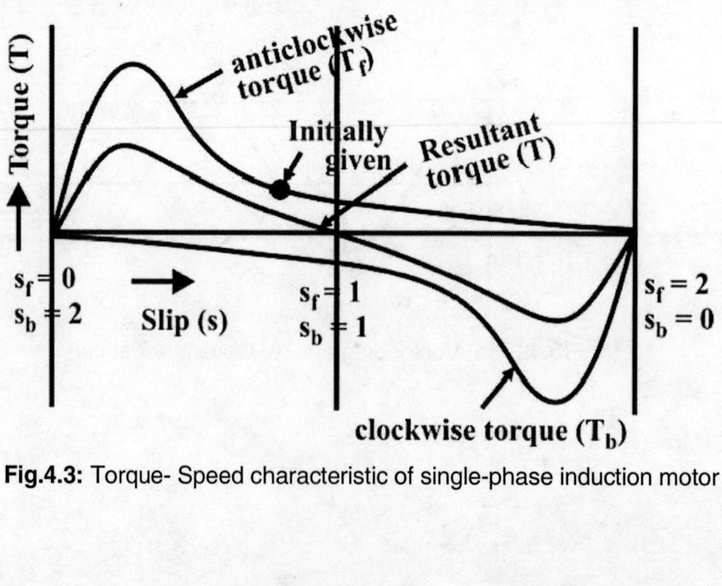

Fig.4.3: Torque- Speed characteristic of single-phase induction motor

At standstill condition, S = 1

Substitute S =1, the starting torque, $T_{st} = 0$

Thus the single phase induction motor is not self-starting.

The torque- speed characteristic of single phase induction motor is shown in Fig.4.3

There are various techniques for starting the single-phase induction motor. Because mechanical methods are not very practical, the motor is temporarily started by turning it into a two-phase motor.

According to the auxiliary device used to start the motor, single-phase induction motors are divided into different categories. They are classified as follows:

a. Split-phase resistance start induction motor
b. Split phase Capacitor start induction motor
c. Capacitor start capacitor run motor
d. Permanent split capacitor (PSC) motor

a. Split-phase resistance start induction motor

In a Split Phase induction motor the status is provided with two parallel winding displaced in space by 90° fig. 4.4 shows the winding windings of the split phase induction motor. The starting winding or auxiliary winding has less turns of smaller diameter to make the winding more resistive (*R=ρl/a*). The running or main winding has thicker turn and a large number of turns to make the winding more inductive.

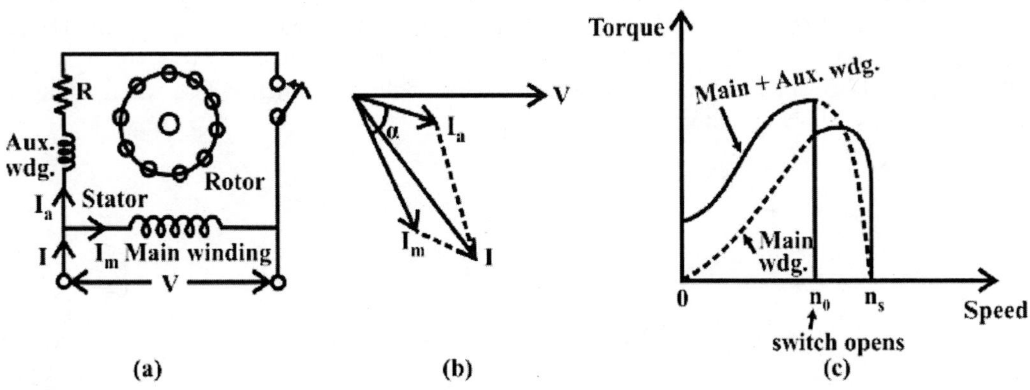

Fig.4.4: Resistance Split-phase induction motor (a) Schematic Diagram, (b) Phasor Diagram and (c) Torque-Speed Characteristics

The vector diagram shows the faith relations of starting winding current I_a and current row main winding I_m. The resistance of the starting winding is increased by the connecting the offer resistance 'R' in series with it.

Hence the current I_a lags behind V by angle of 10^0 the current I_m lags behind the voltage V by a large angle 40^0. The two current I_a and I_m displaced by an angle 30^0 produce a rotating magnetic field (RMF) in the stator. This RMF cuts the rotor conductors. A voltage is induced in the rotor winding this voltage circulate occurring in the rotor conductors. The current carrying rotor conductors experience a force and rotate in the direction of RMF.

After the motor reaches 75% of the rated speed the switch connected in series with the starting winding is disconnected. Thus the starting winding is connected only at the time of starting. The torque developed directly proportional to sinα. Since α is around 30^0 starting torque is less.

$$T = I_a\, I_m \sin\alpha$$

When $\alpha = 90^0$

$$T = T_{MAX}$$

Single phase induction motor starts as a two phase motor runs as single phase motor since α is small it is not used in practice. The torque-speed characteristics of the motor during main and auxiliary winding and only main winding is shown in Fig.4.4(c).

The disadvantages of this motor are Low Starting torque, Noisy. The applications of such motors are Oil burners, Machine tools, Grinders, Washing Machines and Air Blowers.

b. Split phase Capacitor start induction motor

In this motor a capacitor is connected in series with the starting winding as shown in Fig.4.5 (a). Therefore the current in the starting winding will lead the applied voltage. The current through the main winding is lagging the voltage by an angle of 40° as shown in Fig.4.5 (b). The angle 'α' between current I_m and I_a is 80°. The torque is proportional to sinα. Since 'α' is large starting torque is large for this motor. The power factor angle ϕ is when is less when compared with residence split phase motor.

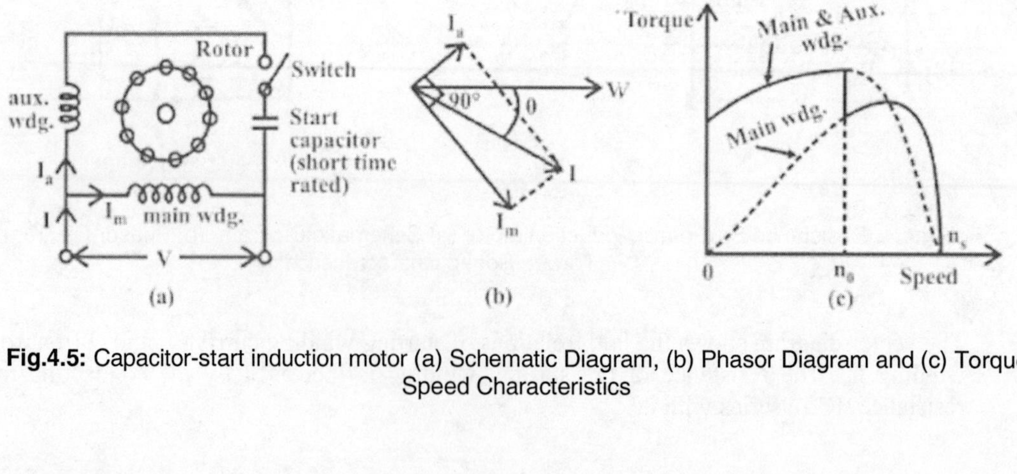

Fig.4.5: Capacitor-start induction motor (a) Schematic Diagram, (b) Phasor Diagram and (c) Torque-Speed Characteristics

The power factor of the capacitor split phase motor is improved the starting torque of this motor is 3 to 4.5 times the full load torque. Once the motor reaches 75% of rated speed the centrifugal switch (CS) opens and disconnected the starting winding from the circuit. The torque-speed characteristics of the motor during main and auxiliary winding and only main winding is shown in Fig.4.5(c). Applications of this motor are such as Pumps, Compressors, Refrigerators, Air conditioners and Washing machines.

c. Capacitor start capacitor run induction motor

In this motor, two capacitors C_s for starting and C_r for running are used as shown in Fig.4.6(a). The first capacitor is designed to be utilised intermittently for beginning. Here, a centrifugal switch is also required. Given that it is utilised for running, the second one must be rated for continuous duty. Figure 4.6 (b and c) depicts the torque-speed characteristics with two windings with different values of capacitors as well as the phasor diagram of two currents in both scenarios.

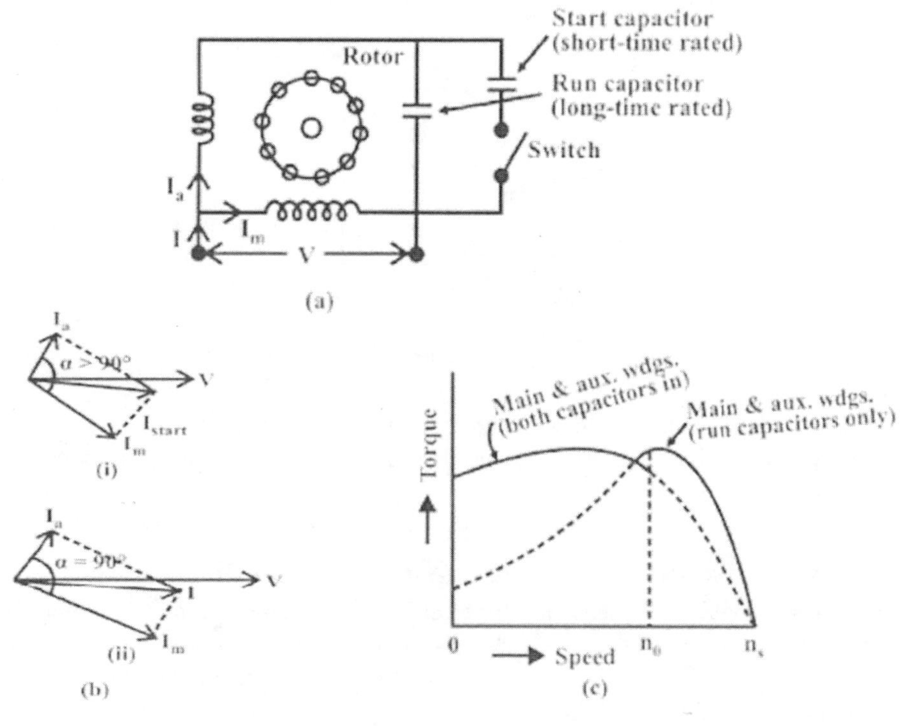

Fig.4.6: Capacitor-start capacitor-run induction motor (a) Schematic Diagram, (b) Phasor Diagram and (c) Torque-Speed Characteristics

When the first case starts, the phasor difference between the two currents is ($\varphi_m + \varphi_\alpha > 90^0$), however when the second case is running, it is 90^0. The second scenario is a balanced two-phase motor that satisfies a fore mentioned requirements as well as having two windings with the equal number of revolutions. Therefore, there is only a forward rotating field and no backward rotating field. The motor's efficiency in this scenario is also fulfilled. The motor operates more effectively in this situation. As a result, using two capacitors improves the motor's performance during startup and operation. Applications for this motor include compressors and refrigerators, among others.

d. Permanent-split capacitor (PSC) motor

In this motor the capacitor connected in series with the auxiliary winding used for starting and running conditions as shown in Fig.4.7. The motor runs continuously with the capacitor and the centrifugal switch is not required.

The motor behaves like a two phase induction motor. The capacitors used in this motor must be designed for continuous duty. The capacitor used in an oil filled capacitor. The noise is reduced because the motor producers are uniform RMF. This motor is costlier than split phase capacitor motor as the auxiliary winding has to be designed for continuous duty.

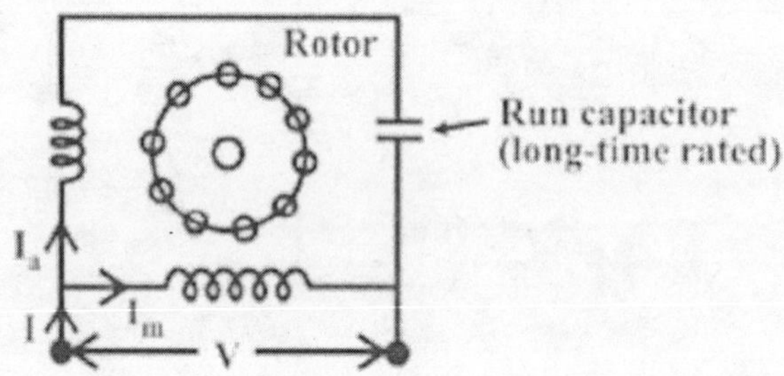

Fig.4.7: Capacitor-run induction motor

The advantages of this motor are increased overload capacity, higher power factor, higher speed and reduced noise. This motor is used in applications such as exhaust fans, blowers, office machine etc.

4.2.2 Maintenance of single-phase induction motors

Regular maintenance is necessary for electric motors in order to prevent failure and increase their lifespan. A minimum of every six months should be set out for maintenance and testing of motors and motor parts. Only then can a motor's life and effectiveness be maintained. We'll discuss three different types of maintenance:

- Preventive maintenance – to prevent operating problems and make sure that the motor continuously provides reliable operation.
- Predictive maintenance- which makes sure that the appropriate maintenance is done at the appropriate time.
- Reactive maintenance- when a failure occurs, the motor is repaired and replaced.

Preventive Maintenance

Regularly performed preventive maintenance on electric motors can assist avoid motor failure and, as a result, unforeseen production interruptions. Preventive maintenance includes some of the most critical components are

- Voltage and current imbalance
- Under voltage and overvoltage
- Motor ventilation
- Humidity and condensation
- Loose connections
- Bearings, Bearing life, Bearing lubrication
- Lubrication type, Lubrication Intervals

Predictive Maintenance

By identifying issues at an early stage and fixing them, predictive maintenance for electric motors aims to lower maintenance costs. A few examples of data that can be used to forecast when the motor needs to be serviced or replaced include observations of the motor's temperature, vibrations, etc. The tests that offer the necessary information regarding the condition of the motor are listed below.

- Cleaning and drying stator windings
- AC high potential phase to ground test and phase-to-phase test
- High potential testing – HIPOT
- Motor temperature
- Surge test
- Bearing considerations
- Insulation-related issues
- Cleaning and drying of the stator windings
- Ground insulation test
- High potential testing,
- Phase-to-phase and ground-to-phase tests for AC and DC high potential systems.
- Thermographic inspection
- Reactive Maintenance

When motors malfunction, it's crucial to analyse the motor to determine what went wrong and why. In most cases, effective preventative maintenance can stop failure. In order to prevent the same failure from occurring elsewhere in the motor or in the entire system, all identical equipment must be inspected if the failure was caused by a weak component or by insufficient maintenance.

4.3 SHADED POLE INDUCTION MOTOR

A single-phase induction motor is not a self-starting motor, as is common knowledge. Changes must be made to the structure for it to get started. Let's examine the shaded pole induction motor's design and operation. A shaded-type induction motor is built similarly to a standard single-phase induction motor, with the exception of the stator pole.

A shaded pole motor has two sections for its stator poles. The motor is referred to as a shaded pole induction motor because one portion of the pole is made up of a short-circuited copper coil called the shaded coil, while the other portion is referred to as the unshaded portion of the pole. The rotor of a three-phase induction motor is constructed similarly to a standard squirrel cage motor. The single-phase, two-pole induction motor with shaded coils on each of the stator poles is shown in Fig.4.8.

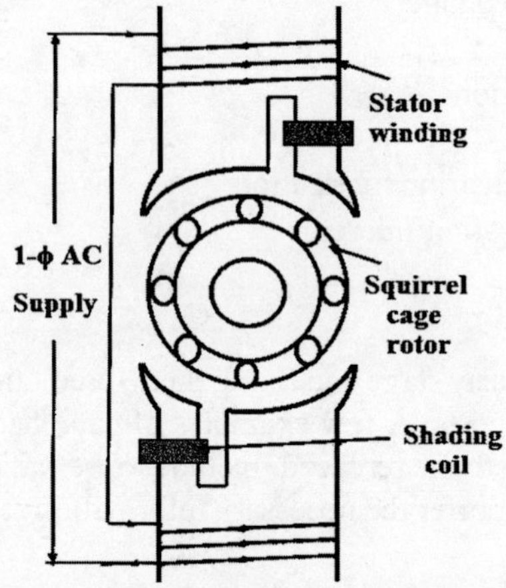

Fig.4.8: Single-phase two-pole induction motor

4.3.1 Working Principle of Shaded Pole Induction Motor

When the stator winding is supplied with a single-phase AC, a magnetic field (Φ_m) is created by the motor's poles, but another field (Φ_s) is produced by coils that are shaded in either the same or the opposite direction.

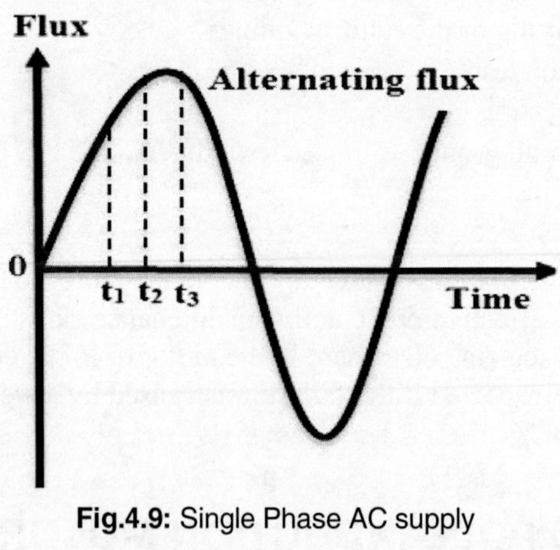

Fig.4.9: Single Phase AC supply

The rotor rotates because of the spinning magnetic field created by the interaction of these fields. Let's take a look at three distinct examples of a positive half cycle of an AC supply at various time instants, t_1, t_2, and t_3.

Case-1 (At instant $t = t_1$):

The waveform in Fig.4.9 shows that at instant $t = t_1$, as may be seen. The flux Φ_m generated by the coil increases as the current in the coil raises. Since the supply is of alternating, the flux's rate of change tends to produce an e.m.f. in the pole's shaded coil portion. Currents will flow through the shaded coil as it is short-circuited, as shown in Fig.4.10 The shaded coil simultaneously develops an additional flux that is directed in the opposite direction from the primary flux, as seen below.

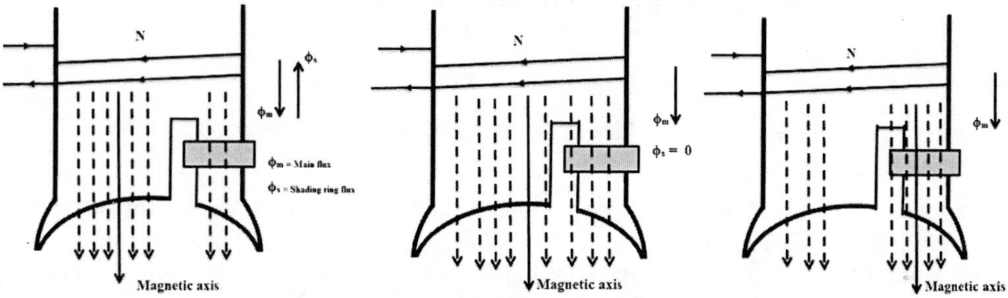

At instant t = t1 At instant t = t2 At instant t = t3

Fig.4.10: Production of torque in a shaded pole induction motor

Thus as a result of the opposition of these two fluxes created by the main winding and shaded coil. The shaded coil's region will experience zero net flux. As a result, the centre of the unshaded area will be where the magnetic axis of the net flow is located.

Case-2 (At instant $t = t_2$)

The peak value of the cycle or current is now available at instant $t = t_2$. No further current will be added or changed in this area. As a result, it slows down the flux's rate of change, reducing the induced e.m.f. in the shade coil in the process. The shaded coil flux (Φ_s) is producing at this stage will be practically not present at all. As a result, the main winding's flux Φ_m will be equally spread around the pole. The magnetic axis of the pole will thus be located in the middle of the whole pole (with shaded and unshaded portions), as seen below.

Case-3 (At instant $t = t_3$)

The rate of change of current will be reducing at this moment. Due to the alteration in flux, there will be an induced e.m.f in the shading coil when the current varies. But in the present scenario, the main winding and the shaded coil will both create fluxes Φ_m and Φ_s that are directed in the same direction. Additionally, the shaded area will have more flux congestion than the unshaded area. As a result, the net magnetic axis of the pole will be located in the middle of the pole's shaded region, as seen below. The negative half cycle likewise features a recurring series of instants as a rotating magnetic field will be created as a result. Where by the motor often starts by itself. This sort of motor produces starting torque that is between 50% and 60% of the full-load torque.

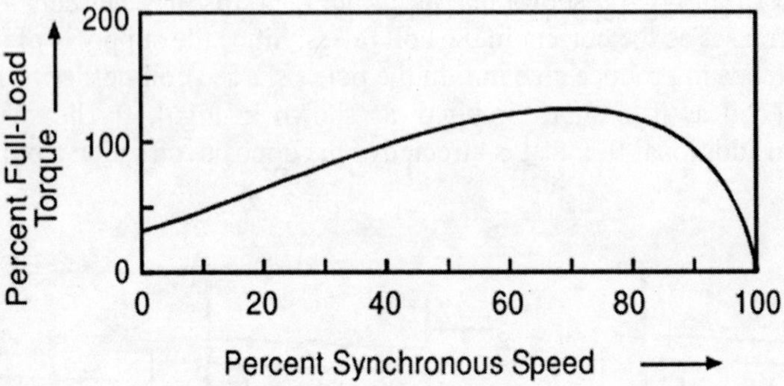

Fig.4.11: Torque-Speed characteristic of shaded pole induction motor

The speed change is really challenging. On each ends of the poles, two sets of shade rings must be installed in order to reverse the rotation's direction. You may accomplish a certain rotational orientation by opening one ring and shutting the other. However, the procedure is costly as well as time-consuming. The typical Torque-Speed characteristic of a shaded pole induction motor is shown in Fig.4.11. This kind of motor's initial torque is just 40 to 50 percent of its full-load torque

Advantages of Shaded Pole Induction Motor

- The design is simple.
- Extremely affordable;
- Rugged.
- A centrifugal switch is not necessary.

Disadvantages of Shaded Pole Induction Motor

- Poor starting torque.
- Copper losses in the shade ring result in extremely low efficiency and a very small over-load capacity.

Applications of Shaded Pole Induction Motor

These have applications in devices with modest starting torque needs, including as tiny fans, toys, hair dryers, movie projectors, and advertising displays.

4.4 REPULSION TYPE MOTOR

Repulsion Motors are a unique type of single phase AC motor that function by repelling poles that are identical to one another. This motor's rotor circuit is shorted by a carbon brush, and the stator is powered by a single phase of AC current.

4.4.1 Construction of Repulsion Motor

The stator, rotor, and commutator brush assembly are the three primary parts of a repulsion motor. A single phase exciting winding comparable to the primary winding of a single phase induction motor is carried by the stator.

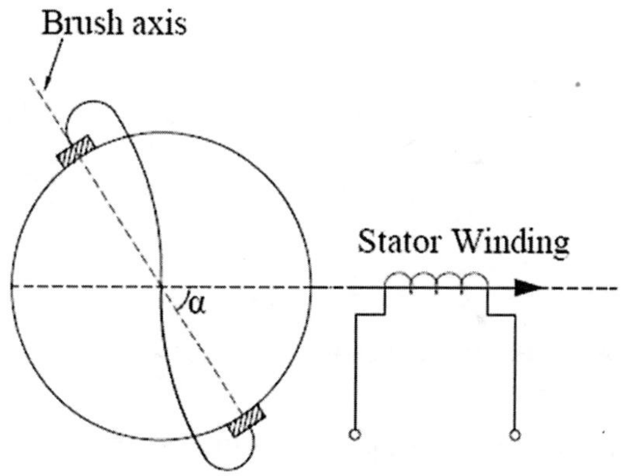

Fig.4.12: Repulsion Motor

Similar to a conventional DC motor, the rotor includes a distributed DC winding that connects to the commutator at one end. The carbon brushes have shorted against one another.

The stator winding in Fig.4.12 above has a single phase AC winding that generates the working m.m.f. in the air gap. It is demonstrated that the rotor's brushes are shorted. Shorting the rotor circuit causes the stator to operate as a transformer, sending power to the rotor.

4.4.2 Working principle of Repulsion Motor

The fundamental theory of how a repulsion motor operates is that "similar poles repel each other." In other words, the two North Poles will oppose one another. Similar to how two south poles repel one another. When the stator winding of a repulsion motor receives a single phase AC supply, it generates a magnetic flux along the direct axis, as indicated by the arrow in the previous figure. When this magnetic flux interacts with the rotor winding, an e.m.f is produced. A rotor current is created as a result of this e.m.f. Due to the commutator assembly, this rotor current then generates a magnetic flux that is directed along the brush axis. An electromagnetic torque is created as a result of the interaction of the fluxes produced by the stator and rotor.

The angle between the stator-produced field and brush axis in Fig.4.13(a) above is 90°. This indicates that the brush axis and the direct axis are in quadrature. The stator and rotor windings won't induce each other under these circumstances. As a result, no e.m.f and hence no rotor current are generated. No electromagnetic torque is thus produced. As a result, the motor won't operate when α=90°. This state is analogous to an open circuit transformer since the zero rotor m.m.f has no effect on the flux produced by the stator. This is the basis for the terms open-circuit, no-load, high impedance, and neutral used to describe the brush position at α = 90°. Consider the example in which α=0°, as illustrated in Fig.4.13(b).

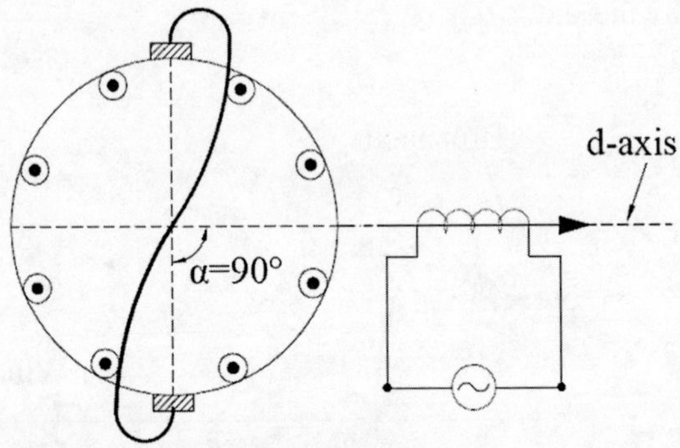

Fig.4.13 (a): Angle between the stator field and brush axis, ▯ = 90°

Peak e.m.f is induced throughout the brushes in this case. This is due to the fact that the magnetic fluxes of the rotor and stator coincide, resulting in perfect mutual coupling. Because the electromagnetic torque T is represented as

$$T_e = k \text{ (Stator Field Strength) (Rotor Field Strength) } \sin\alpha$$

where is k is a constant.

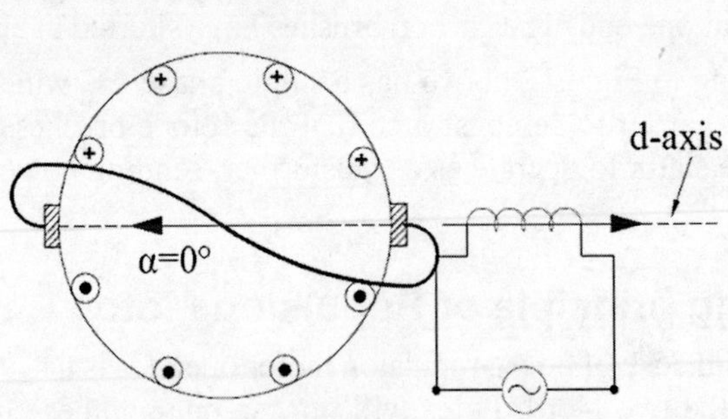

Fig.4.13 (b): Angle between the stator field and brush axis, ▯ = 0°

As α = 0°, no electromagnetic torque is produced. Consequently, in a repulsion motor, no electromagnetic torque is produced when the angle between the rotor and stator magnetic flux axes is either zero or 90 degrees.

However, as illustrated in Fig.4.14, the brush axis is located somewhere between $\alpha = 0°$ and $90°$. If the flux produced by stator is considered to be oriented from A to B, then the rotor flux produced must contain a component in the opposite direction. This is simply due to Lenz's Law. As a result, the rotor flux is directed from C to D. It should be noted that it cannot be directed from D to C because this would result in a flux component directed from A to B, which would violate Lenz's Law.

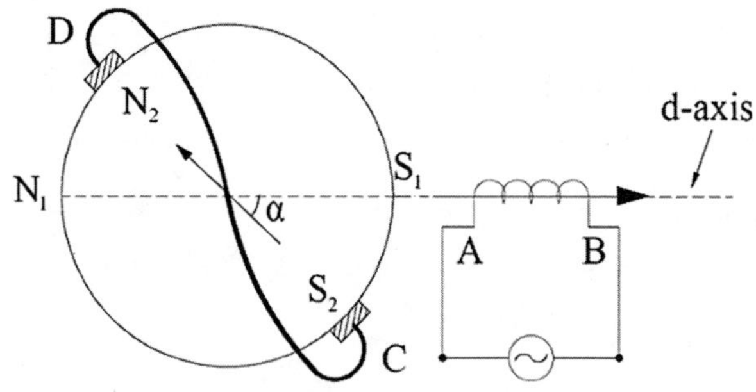

Fig.4.14: Brush axis between □ = 0° and □ = 90°

As the stator flux is directed from A to B, the South Pole (S_1) is formed at A. Similarly, at C, the South Pole (S_2) is generated. S_1 will repel S_2 because comparable poles repel each other. Because of this repulsion between like poles, the motor will revolve clockwise. This is why this motor is known as the Repulsion Motor. The following diagram and explanation demonstrate that the rotational direction of a repulsion motor may be reversed by simply shifting the brush axis to the opposite side of the filed winding (stator winding).

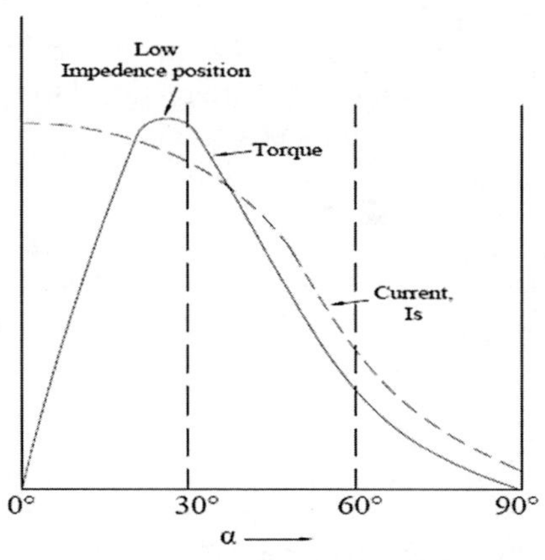

Fig.4.15: Variation of current and torque with respect to different positions of brush

Figure 4.15 depicts the changes of current and torque with regard to different brush positioning. The following points should be observed from aforementioned curve:

- Rotor current is maximal when the direct and brush axes coincide.
- When the brush is in quadrature with the direct axis, the rotor current is zero.

Maximum torque is obtained in a repulsion motor when the stator and rotor field axes are 45° apart. Repulsion motors are used in Hoists, Machines in Textile, Printing presses, Pumps & Fans, mixing machines, machine tools, petrol pumps and drive compressors.

Advantage of a Repulsion Motor

A repulsion motor has many advantages, including high starting torque, low starting current, and a broad range of speed control with smooth speed variation. The following are a few of the disadvantages of the repulsion motor. The power factor is substantially lower at lower speeds. Due to heat production and arcing at the brush assembly, brushes and commutators quickly exhaust.

Subsequently, it all comes down to the repulsion motor. Most commutator motors are limited to 1500 V because excessive voltages provide a risk of arcing across them. Because the rotor circuit is not electrically coupled to the power supply, these types of motors are used where high voltages are necessary.

4.5 AC SERIES MOTOR

A conventional DC series motor will always run in the same direction, regardless of supply polarity. The torque direction is determined by the respective positions in space of flux and armature current. When the line terminals are flipped, the field and armature currents are reversed, but the torque direction stays unchanged. As a result, the motor keep rotating in similar direction. When a standard DC series motor connects to an alternating current supply, both armature and field currents reverse at a time, producing unidirectional torque in the motor. Consider a two-pole motor with the alternating current in its positive half; the polarity of the field poles and currents flowing through the armature conductors will be as shown in Fig.4.16.

The armature conductors circulate positive currents under N-pole and negative currents under S-pole. Using Fleming's left-hand rule, we can see that the torque created in the armature will try to rotate anti-clockwise.

The alternating current then passes through the negative half cycle, as shown in Fig.4.17. The current flowing through the field winding and armature will now change as well. Because of the uniform torque created by the two parts of the cycle, the armature will tend to rotate in the same direction.

The series motor may operate on both DC and AC supplies as a result. Because of the following factors, a DC series motor's functioning from an AC source is not satisfactory:

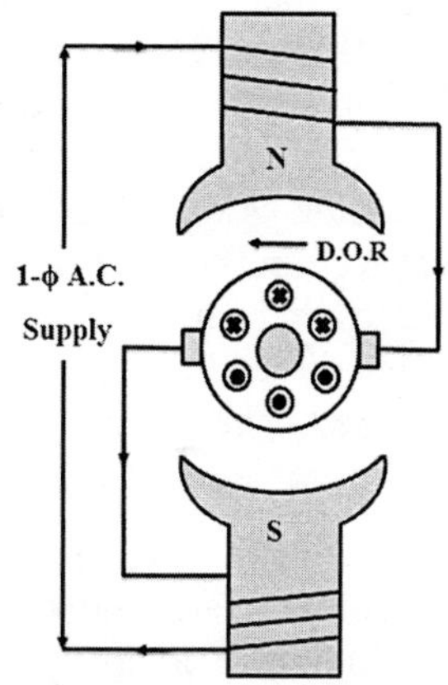

Fig.4.16: Polarity of the field poles in the ac series motor first half

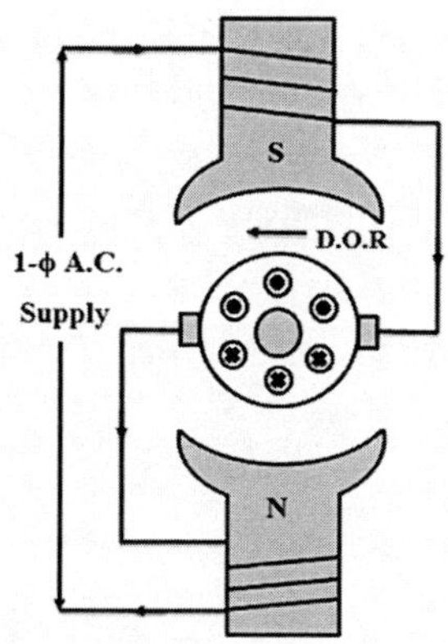

Fig.4.17: Polarity of the Field Poles in the AC Series Motor Second Half Cycle

The efficiency is inappropriate. This is brought on by the alternating flux's increase in core wastes.

- As the source provided is alternating, the reactance of the armature and field windings increases, causing the device to run at a low power factor.
- There will be a lot of sparking at the brushes. Commutation issues are made worse by the voltage the transformer performance generates in the coil that is undergoing commutation.

Construction of AC Series Motor

When a DC series motor is referred to as an AC series motor, several modifications are required to ensure that it operates satisfactorily on the AC source. To reduce eddy current wastes, fully laminated yokes and poles should be used. By lowering the field reactance and armature reactance, the power coefficient can be increased. The field coil is designed with fewer turns in order to reduce the field reactance. Additionally, a lower pole flux reduces the EMF of the transformer in the commutating winding. The motor needs to have a lot of poles, each of which provides less flux overall. The field flux would likewise be reduced if the field coil's turns were reduced. The armature turns must be increased proportionally in order to maintain the torque fixed and smooth on the shaft. This may enhance the armature's responsiveness and reactance. The armature reactance should be reduced as much as possible using compensating coil. Compensation improves commutation as well. Since the compensating coil's flux is the opposite of the armatures, it effectively cancels out the response caused by the armature. To avoid the short circuit problem, single-turn brushes and narrower coils are used in the armature windings. The extremely constrained air volume allows for the application of smaller field rotations per pole. The source's frequency is decreased. Therefore, a suitable commutation is attainable at lower frequencies since the transformer EMF is related to frequency.

Since the field and armature circuit reactance now directly affects power factor, we can lower reactance by reducing the field winding's number of turns. However, there is one issue: when the number of revolutions is decreased, the field m.m.f will also decrease, which will result in a decrease in the air gap flux. Overall, this causes the motor speed to increase but the motor torque to decrease, which is not what is expected. Now, how can we solve this issue? The adoption of compensatory winding is the solution to this issue. There are two different of motors based on how the compensatory winding is used, and they are as follows:

- Motors of the conductively compensated type.
- Motors of the inductively compensated type.

Conductively Compensated Type of Motors

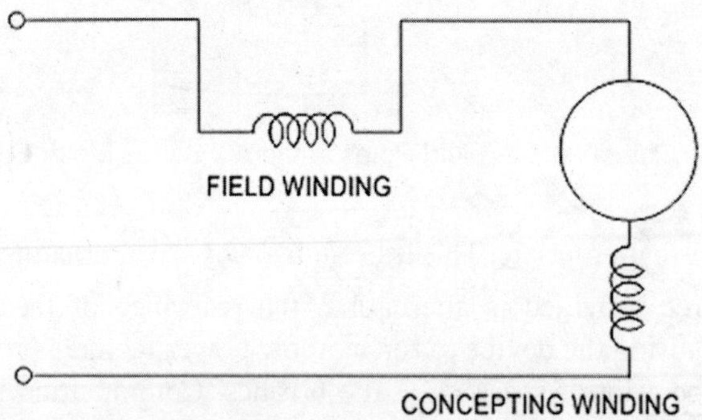

Fig.4.18: Conductively Compensated type of AC series motor

The circuit diagram for the conductively compensated type of motors is shown in Fig.4.18, which is provided below. The compensatory winding and armature circuit are linked in series in this type of motor. The stator slots receive the winding. The primary field axis and the axis of the compensatory winding are electrically 90 degrees apart.

Inductively Compensated Type of Motors

The circuit diagram for the inductively compensated type of motors is shown in Figure 4.19, which is provided below. The compensatory winding in this type of motor is not connected to the motor's armature circuit. The armature winding will function as the transformer's primary winding in this instance, while the compensating winding will serve as the secondary winding. The compensatory winding's current will be in phase opposition to the armature windings current.

The single phase AC series motor complete schematic diagram, including inter pole and compensatory winding adjustments, is shown in Fig.4.20.

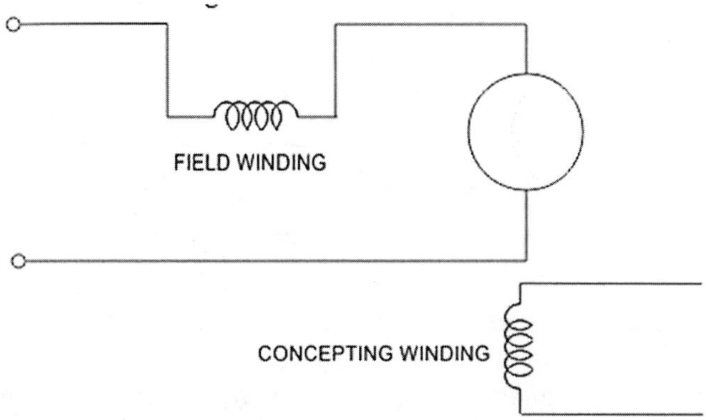

Fig.4.19: Inductively compensated type of ac series motor

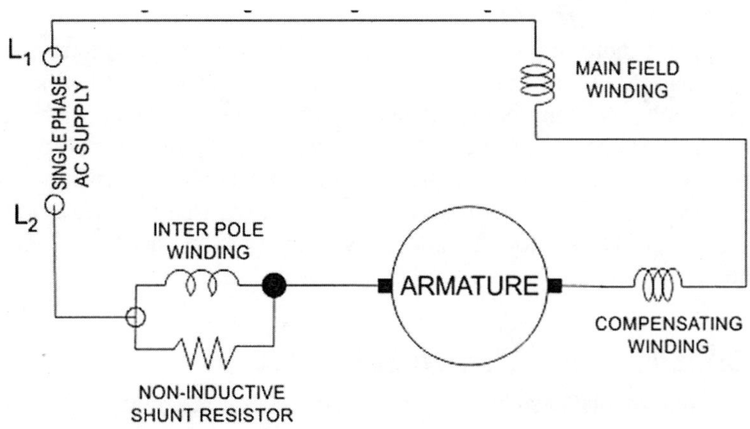

Fig.4.20: AC series motor with interpoles and compensating winding

A solid-state device is the most effective way to control the speed of this sort of motor. The benefit of having compensating winding has already been covered. Let's talk about the interpoles purpose. The inter poles' primary purpose is to increase the motor's performance in terms of improved efficiency and a greater output from the predetermined size of the armature core. In order to lower the series field inductance, we have taken a very large reactive voltage drop in comparison to either the armature or the compensatory field.

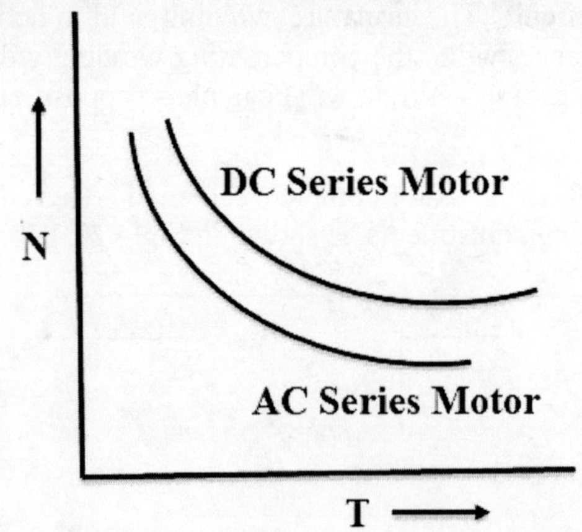

Fig.4.21: Torque-Speed characteristic of the ac series motor

As depicted in a fore mentioned Fig.4.20, the winding of the inter pole circuit is linked in parallel with the non-inductive shunt. Numerous appliances, including portable drills, hair dryers, table fans, and culinary appliances use the motor. Torque speed characteristics of ac series motor is given in Fig.4.21.

4.6 UNIVERSAL MOTOR

A special type of motor known as a "universal motor" is designed to run on either a DC supply or a single phase of AC power. These motors frequently have field windings and series wound armatures, which produces a high starting torque. As a result, the machinery that universal motors are intended to power typically comes with them already installed. Most universal motors are designed to work at speeds greater than 3500 RPM. They run more slowly under an AC source than they do with a DC supply of the same voltage because reactance voltage loss occurs in AC but not in DC. Universal motors come in two fundamental varieties: (i) compensated type, and (ii) uncompensated type.

4.6.1 Construction of Universal Motor

The architecture of a DC machine and a universal motor are remarkably similar. It is made up of a stator, which is positioned atop field poles. Weaved field coils are attached to the field poles. But the armature and the stator field circuit, along with the complete magnetic path, are laminated. Lamination is crucial to lowering the AC's eddy currents. Brushes are positioned on the commutator, and the rotary armature is wound with either straight or skewed slots. The commutation on AC is less efficient than that on DC because of the current generated in the armature coils. As a result, the used brushes have a high resistance. Fig.4.22 depicts the cross-sectional view of Universal Motor, and Fig.4.23 depicts the sectional view of non-compensated type universal motor.

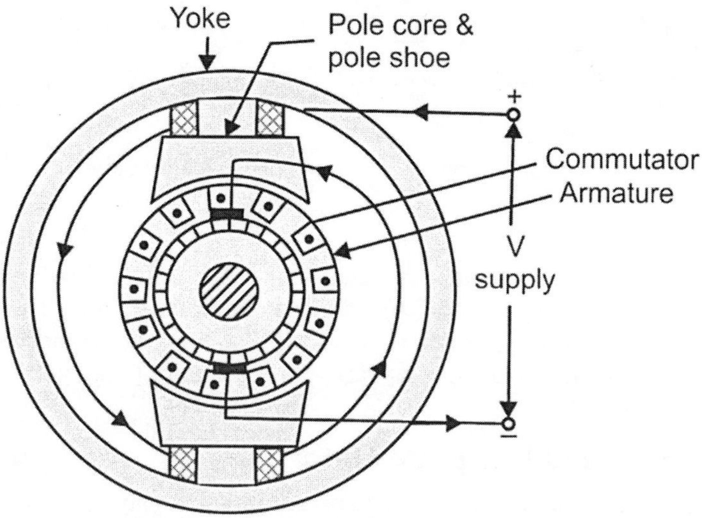

Fig.4.22: Sectional view of Universal Motor

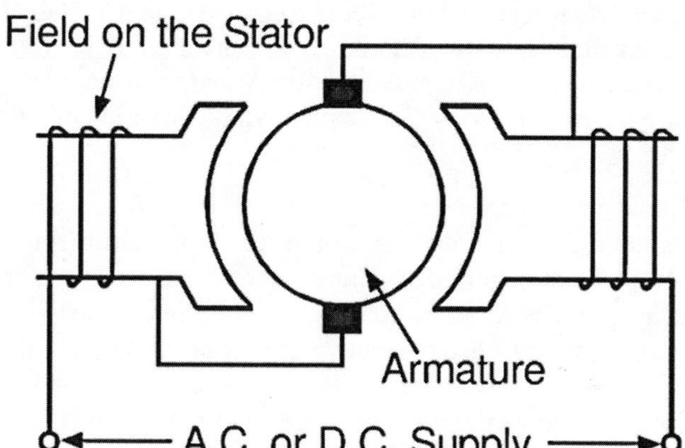

Fig.4.23: Cross-sectional view of non-compensated type universal motor

There are two different kinds of universal motors in use: compensated and non-compensated. As seen in Fig.4.24(a), the non-compensated motor is typically constructed with concentrated or prominent poles.

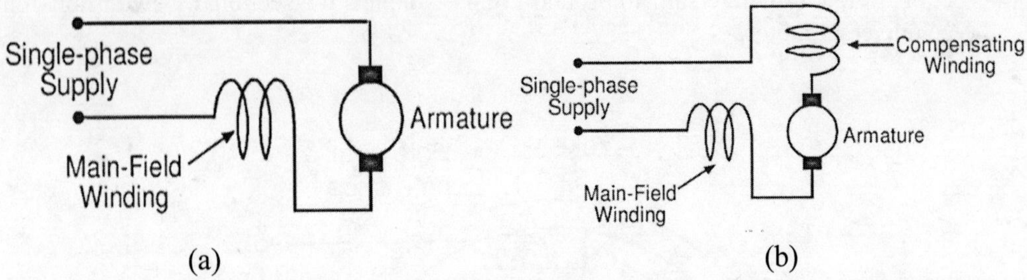

Fig.4.24: shows the connections for two types of universal motors: (a) non-compensated and (b) compensated.

The compensated motor, on the other hand, has a scattered field winding. Therefore, the stator of a motor like this resembles a split phase induction motor. Both of these motor designs have a coiled armature that resembles a tiny DC motor. The connection schematic for a universal motor is shown in Fig.4.24. Non-compensated type in (a), compensated type in (b).

4.6.2 Working of Universal Motor

A universal motor can run on a single phase AC or DC supply. The universal motor functions as a DC series motor when it is supplied with a DC source. An electromagnetic field is created when electricity travels through the field winding. The conductors of the armature likewise carry the same current. A current-carrying conductor encounters a mechanical force when it is exposed to an electromagnetic field. The rotor begins to revolve as a result of this mechanical force, or torque. The Fleming left hand rule determines the direction of this force. It still generates a single-direction torque when supplied with AC power. Armature winding and field winding are in phase because they are coupled in series. Consequently, the direction of current changes as the polarity of AC fluctuates.

In order to maintain the same direction of force on armature conductors, the direction of the magnetic field and the direction of armature current reverse. Therefore, universal motors operate on the same principles as DC series motors, whether they are powered by an AC or DC supply. A universal motor's torque-speed characteristics are comparable to those of a DC series motor. A universal motor's speed is low under full load and extremely high under no load. Gear trains are typically utilised to achieve the desired speed on the demanded load. The graphic displays the speed/load characteristics for both AC and DC supplies.

Applications of Universal Motor

- Home equipment like hoover cleaners, drink and food mixers, domestic sewing machines, etc. all employ universal motors.
- Portable drills, blenders, and other appliances employ universal motors with greater ratings.

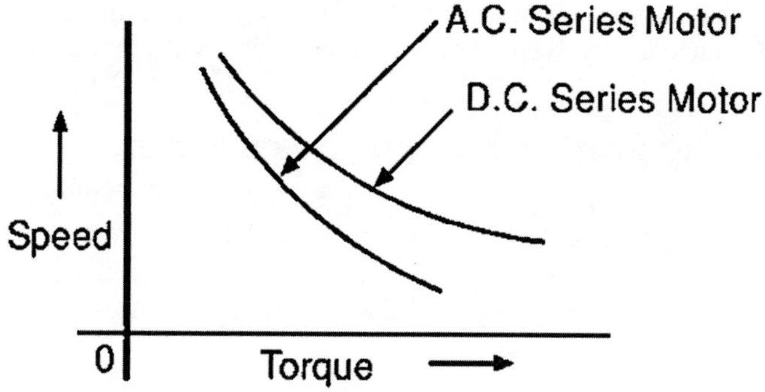

Fig.4.25: Torque-Speed characteristics of a universal motor

4.7 HYSTERESIS MOTOR

The synchronous Hysteresis Motor has a cylindrical rotor, uses non-projected poles, and does not require any DC excitation of the rotor. It is a single-phase motor with a ferromagnetic material-covered rotor. For the purpose of creating a rotating magnetic field, the stator is wound with both primary and auxiliary windings. Some hysteresis motor designs additionally allow for shaded pole stator options.

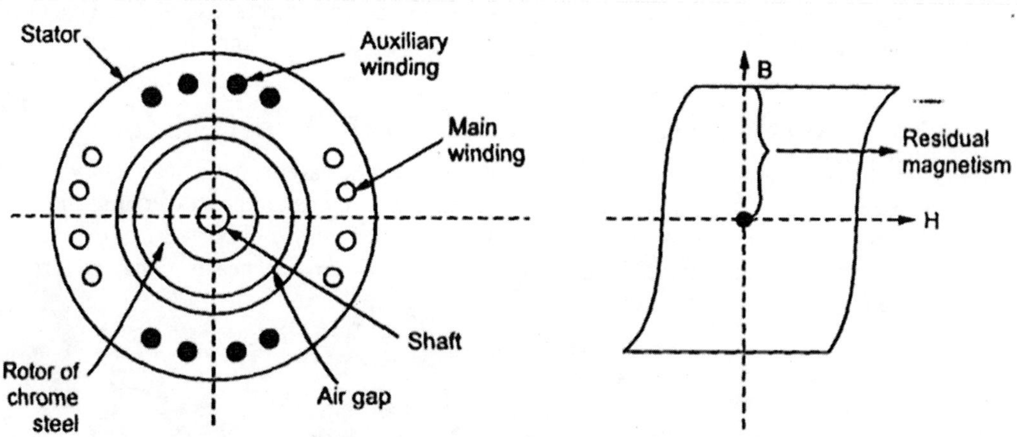

Fig.4.26: Construction of hysteresis motor & hysteresis loop for rotor

The stator of a hysteresis motor is designed in such a way as to create synchronising rotating field from single phase supply. The rotor has a smooth, cylindrical design constructed of high retentivity hard magnetic materials like chrome steel or alnico. For this, a material with a large hysteresis loop area must be chosen. There are no teeth or windings on the rotor. The hysteresis motor's rotor has a great resistance to reduction. Hysteresis motor construction and the type of hysteresis loop needed for the rotor material is depicted in Fig.4.26

4.7.1 Working Principle of Hysteresis Motor

A hysteresis motor initially operates as a single phase induction motor, but as it runs, it transforms into a synchronous motor. A revolving magnetic field is created when a stator is powered by a single phase AC supply. To sustain the revolving magnetic field, both the main and auxiliary windings must be continually supplied during start up and operation. This field causes the rotor to produce poles.

Due to the rotor material's predominance of the hysteresis phenomena, the rotor pole axis is offset from the axis of the revolving magnetic field. Rotor poles are drawn to the moving stator field poles as a result. As a result, a torque known as hysteresis torque is applied to the rotor. At all speeds, this torque is constant when the axis of the stator.

Hysteresis torque and torque brought on by induced eddy currents in the rotor cause the rotor to initially begin rotating. The stator pushes the rotor into synchronism whenever the speed is close to that of synchronous motion. In this scenario, the torque caused by eddy currents also disappears when the relative motion between the stator field and rotor diminishes. All that is present is hysteresis torque, which keeps the rotor spinning at synchronous speed. The constant magnetic locking between the stator and rotor is ensured by the strong retentivity. The Hysteresis Motor either rotates at synchronous speed or not at all because of the magnetic locking concept.

4.7.2 Torque Speed characteristics of Hysteresis Motor

In this hysteresis motor, the running and starting torques are nearly equal. The direction of the stator, which primarily carries the two windings, can be changed by switching the terminals of either the main winding or the auxiliary winding. As can be observed from the torque speed characteristics in the Fig.4.27, the torque at start-up remains nearly constant throughout the operation of the hysteresis motor.

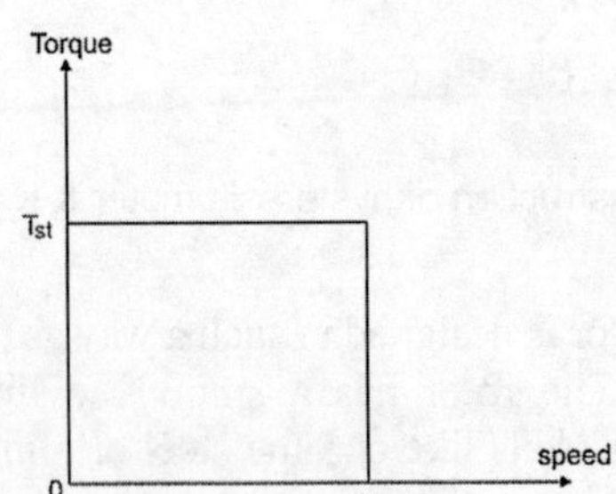

Fig.4.27: Torque Speed characteristics of Hysteresis Motor

The benefits of a hysteresis motor include

- Because the rotor has no teeth or winding, there are no mechanical vibrations.
- As a result, the operation is quiet and noiseless.
- High inertia loads can be accelerated.
- Multispeed operation is possible by using a gear train.

The drawbacks of the hysteresis motor include

- Low power factor, low torque, and low efficiency
- Comes in a very small size
- Poor performance

Applications of Hysteresis Motor

Because it operates quietly, hysteresis motors are employed in sound recording instruments, high-quality record players, electric clocks, tele printers and timing devices.

UNIT SUMMARY

The aim of this chapter is to understand the construction and working principle of the single phase induction motor. The principle of operation and double field revolving theory of the motor conveys clearly why induction motor is called asynchronous machine. Also, discussed about the different starting methods and how it helps in starting the of single phase induction motor. The significance of repulsion motor, universal motor, ac series motor and hysteresis motor were discussed in detail.

EXERCISES

1. A double field revolving theory is applied to a three-phase induction motor with the following parameters: P=4, f=50 Hz, s=0.02. Calculate the synchronous speed, rotor speed, and slip.

2. For a capacitor start capacitor run motor, the capacitance of the start winding is $20\mu F$, and the capacitance of the run winding is $30\mu F$. If the motor operates at 1750 rpm and has a torque of $25Nm$, calculate the slip and the frequency of the rotor.

3. A series motor is required for a specific application with a load torque of $15Nm$ and a speed of 15001500 rpm. Determine the suitable series motor based on torque-speed characteristics.

4. A single-phase induction motor is experiencing high temperature during operation. If the motor is rated at $1HP$ and operates at $230V$, $50Hz$, and $8A$, calculate the power factor and suggest measures to reduce temperature and improve efficiency.

Multiple Choice Questions

1. Single-phase induction motor are
   a. Self-starting
   b. Not self-starting
   c. Self-starting with the help of an auxiliary winding
   d. Need another motor

2. If a single-phase induction motor is running at N rpm and the synchronous speed is N_S. If its slip with respect to forward field is *s*, what is the slip with respect to the backward field?
   a. s
   b. - s
   c. (1 - s)
   d. (2 - s)

3. The rotating magnetic field is produced by two windings displaced by 90 electrical degrees. This is the principle of
   a. Phase sequence
   b. Phase splitting
   c. Phase timing
   d. Phase shifting

4. The torque developed by a single-phase induction motor drop to zero at
   a. Synchronous speed
   b. Speed slightly above synchronous speed
   c. Speed slightly below synchronous speed
   d. Rated speed

5. The no load current of a single-phase induction motor is around % of full load current
   a. 10
   b. 20
   c. 40
   d. 80

6. The power factor at which singie phase induction motors usually operate is
   a. 0.7 lag
   b. 0.8 lag
   c. 0.7 lead
   d. Unity

7. Single phase induction motors are made self-starting by
   a. Increasing rotor resistance

b. Using an external starting device
c. Using auxiliary winding
d. Any of the above methods

8. The two windings provided on the stator of a single-phase induction motor, one main winding and the other auxiliary winding are connected
   a. In parallel
   b. In series
   c. Either in series or in parallel
   d. Through inductive coupling

9. The stator winding of a single-phase induction motor is splitted into two parts to
   a. Improve efficiency
   b. Improve power factor
   c. Develop starting torque
   d. Increase speed

10. In a single-phase induction motor
    a. Both the main and auxiliary windings are placed on stator
    b. Both the main and auxiliary windings are placed on rotor
    c. Main winding is placed on stator and auxiliary winding on rotor
    d. Auxiliary winding is placed on stator and main winding on rotor

11. In a split phase induction motor, the ratio of number of turns of auxiliary winding to the main winding is
    a. 1
    b. < 1
    c. > 1
    d. 2

12. Why a centrifugal switches is are used in a single-phase induction motor?
    a. To protect from overloading
    b. To improve the starting performance of the motor
    c. To cut off the starting winding at an appropriate instant
    d. To cut in the capacitor during running conditions.

13. Centrifugal switch fitted on the rotor will work when
    a. Rotor speed reaches its rated conditions
    b. Rotor speed exceeds 70 per cent of its rated value
    c. Rotor speed exceeds synchronous speed
    d. Rotor speed exceeds 40 per cent of its rated value

14. The torque speed characteristic of two-phase induction motor is largely affected by
    a. Supply Voltage
    b. Speed of the motor
    c. X/R ratio
    d. Supply frequency

15. The direction of rotation of a split phase induction motor can be reversed by reversing the connections to the supply of
    a. Auxiliary winding only
    b. Main winding only
    c. Either (a) or (b)
    d. Both (a) and (b) simultaneously

16. The capacitor in a capacitor start induction run AC motor is connected in series with
    a. Starting winding
    b. Running winding
    c. Rotor winding
    d. Compensating winding

17. The shaded pole motor is used for
    a. High starting torque
    b. Low starting torque
    c. Medium starting torque
    d. Very high starting torque

18. In a shaded pole motor, shading coils are used to
    a. Reduce winding losses
    b. Reduce friction losses
    c. Produce rotating magnetic field
    d. Protect against sparking

19. In a shaded pole induction motor, locked rotor current is ____full load current
    a. Less than
    b. Equal to
    c. Slightly more than
    d. Several times the

20. The direction of rotation of a shade pole induction motor
    a. Cannot be reversed unless there is a provision for shifting of shading coil from one half to the other half of the pole
    b. Can be reversed by interchanging
    c. Can be reversed by interchanging the supply terminals
    d. Can be reversed by open circuiting the shading rings

21. The repulsion motor starts and runs as a
    a. Split-phase motor
    b. Capacitor-start motor
    c. Repulsion motor
    d. Compound motor

22. A repulsion motor runs as an induction motor when the
    a. Commutator segments are short-circuited
    b. Brushes are shifted to a neutral plane
    c. Shorting devices are disconnected
    d. Stator connections are reversed

23. The commutator in a repulsion motor provided in
    a. Armature windings
    b. Stator windings
    c. Short-circuiting device
    d. Starting-winding

24. Hysteresis motor is a
    a. Synchronous induction motor
    b. Single phase induction motor
    c. Single phase synchronous motor without any salient pole and without DC excitation
    d. Single phase synchronous motor with salient poles and without DC excitation

25. Hysteresis motor operates on the principle of
    a. Hysteresis loss
    b. Eddy current loss
    c. Electromagnetic induction
    d. Magnetization rotor

26. The rotor of a hysteresis motor is built up of
    a. Cast iron sheet
    b. Hardened steel rings
    c. Think silicon steel laminations

d. Thin metal laminations

27. A hysteresis motor
    a. Is a self-starting motor
    b. Does not need DC excitation
    c. Does not need Pony motor
    d. All of the above

28. The main reason for using a hysteresis motor for high quality tape recorders and record players is that
    a. Its speed is constant (synchronous)
    b. It develops extremely steady torque
    c. It requires to centrifugal switch
    d. Its operations do not affect by mechanical vibrations

29. The electric motor generally used in household food mixers is
    a. Universal motor
    b. Shaded pole motor
    c. Capacitor start motor
    d. None of the above

30. Which motor has no teeth or winding?
    a. Split phase motor
    b. Reluctance motor
    c. Hysteresis motor
    d. Universal motor

31. Motor which can be produce uniform torque from standstill to synchronous speeds is
    a. Universal motor
    b. Stepper motor
    c. Reluctance motor
    d. Hysteresis motor

Answers of Multiple Choice Questions

1. c. Self-starting with the help of an auxiliary winding
2. d (2 - s)
3. b Phase splitting
4. c. A speed slightly below synchronous speed
5. c. 40
6. a. 0.7 lag

7. c. Providing an additional winding on the stator called the auxiliary winding

8. a. In parallel
9. c. Develop starting torque
10. a. Both the main and auxiliary windings are placed on stator
11. b. Less than one
12. c. To cut off the starting winding at an appropriate instant
13. b. Rotor speed exceeds 70 per cent of its rated value
14. c. X/R ratio
15. c. Either (a) or (b)
16. a. Starting winding
17. b. low starting torque
18. c. produce rotating magnetic field
19. c. slightly more than
20. a. Cannot be reversed unless there is a provision for shifting of shading coil from one half to the other half of the pole
21. c. repulsion motor
22. a. commutator segments are short-circuited
23. a. armature windings
24. c. Single phase synchronous motor without any salient pole and without dc excitation
25. a. Hysteresis loss
26. b. A group of specially hardened steel rings
27. d. All of the above
28. b. It develops extremely steady torque
29. a. Universal motor
30. c. Hysteresis motor
31. d. Hysteresis motor

Short and Long Answer Type Questions

Long Answer Questions with Answers

1. **Why is the fundamental approach for starting a single phase induction motor not recommended?**

   A single-phase motor's rotor will continue to spin in the same direction if it is mechanically rotated in that direction. In fact, the rotor accelerates swiftly until it reaches a speed that is just a little bit slower than the synchronous speed. The motor will continue to rotate after it reaches this speed even though the stator winding is only receiving single-phase electricity. For large motors, this way of starting is typically not practical. It also cannot be used to power a motor that is situated in an inconvenient location.

2. **Given the Double-Field Revolving Theory, why can't a single phase induction motor start?**

The double-field revolving theory is put up as a solution to the paradox of zero initial torque and subsequent rotational torque. This hypothesis is based on the observation that an alternating sinusoidal flux can be represented by two rotating fluxes rotating at synchronous speed in opposite directions, each flux being equal to one-half of the highest value of the alternating flux (i.e., m/2).

Therefore, two related fields revolving at synchronous speed in opposing directions and with a half of the amplitude of an alternating field can replace it. The resulting vector of two rotating flux vectors is an oscillating stationary vector along the X-axis that changes length over time.

3. **What is the rotor's position when it is stationary?**

Consider the scenario when the stator winding is coupled to a single-phase supply and the rotor is immobile. The alternating flux generated by the stator winding can be represented as the total of two rotating fluxes 1 and 2, each of which is revolving at synchronous speed ($N_s = 120$ f/P) in the opposite direction and is equal to one half of the maximum value of alternating flux.

At a complete stop, these two torques are equal and in opposition, resulting in a zero net torque development. Consequently, a single-phase induction motor cannot start on its own.

4. **What causes a motor to move from a resting to operating state?**

Since both circuits' impedances are equal when the system is at rest, $s = 1$. Rotor currents are therefore equivalent, or $I_{2f} = I_{2b}$. The rotor current I_{2b} is larger (and also has a lower power factor) than the rotor current I_{2f} when the rotor is rotating due to the unequal impedances of the two rotor circuits. The backward rotating flux will be decreased as a result of their m.m.f., which is in opposition to the stator m.m.f.

As a result, the driving torque increases as speed increases while the opposing torque reduces as the backward flux drops. The motor immediately picks up speed to reach final speed.

5. **How to make Single-Phase Induction Motor Self-Starting?**

The single-phase induction motor is not self- starting and it is undesirable to resort to mechanical spinning of the shaft or pulling a belt to start it. To make a single-phase induction motor self-starting, we should somehow produce a revolving stator magnetic field. This may be achieved by converting a single-phase supply into two-phase supply through the use of an additional winding. When the motor attains sufficient speed, the starting means (i.e., additional winding) may be removed depending upon the type of the motor.

6. Name the types of single phase motors with respect to the method employed to make them self-starting?

(i) Split-phase motors-started by two phase motor action through the use of an auxiliary or starting winding.

(ii) Capacitor motors-started by two-phase motor action through the use of an auxiliary winding and a capacitor.

(iii) Shaded-pole motors-started by the motion of the magnetic field produced by means of a shading coil around a portion of the pole structure.

7. What is the major cause of noisy operation of the single phase induction motor?

If the two windings arc displaced 90° electrical but produce fields that are not equal and that are not 90° apart in time, the resultant field is still rotating but is not constant in magnitude. One effect of this non uniform rotating field is the production of a torque that is non-uniform and that, therefore, causes noisy operation of the motor. Since 2-phase operation ceases once the motor is started, the operation of the motor then becomes smooth.

8. Describe the principal of operation of Split-Phase Induction Motor?

The stator of a split-phase induction motor is provided with an auxiliary or starting winding S in addition to the main or running winding M. The starting winding is located 90° electrical from the main winding and operates only during the brief period when the motor starts up. The two windings are so resigned that the starting winding S has a high resistance and relatively small reactance while the main winding M has relatively low resistance and large reactance as shown in the schematic connections in Fig. Consequently, the currents flowing in the two windings have reasonable phase difference c (25° to 30°) as shown in the phasor diagram in Fig.

9. Usually where split phase induction motors are used?

These motors are suitable for driving things like (a) fans, (b) washing machines, (c) oil burners, (d) small machine tools, etc. when a moderate starting torque is needed.

These motors typically range in power from 60 W to 250 W.

10. What is Shaded Pole Motor?

The shaded pole induction motor is a simple single-phase induction motor that is self-starting with one of the poles shaded by a copper ring. A copper ring can also be called a shaded ring, where it serves as a secondary winding motor. There is only one direction of rotation, and it is not possible to reverse the direction. There are very low power factor and high power induction losses in this motor. The induced starting torque in the motor is very low. Because of these reasons, its efficiency is low. Due to this, it has a low power rating. The motor is also a salient pole split phase motor.

11. Explain the Construction of Shaded Pole Induction Motor.

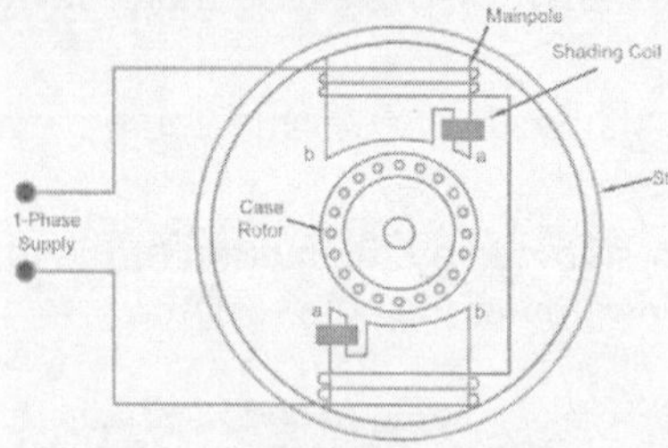

According to the basic construction, there are two poles. This motor consists of a stator and a cage-type rotor. Stators have projected poles, also called main poles. A main winding is formed by the supply winding on the main poles. The poles in this motor are unequally divided into two halves, with the smaller half carrying a copper band. On the smaller part is a copper ring with a single turn. Alternatively, this ring is called a shading coil. Shading poles are fitted with shading coils on the main poles.

12. Explain the working principle of shaded pole induction motor.

A flux is induced in the pole's main part when power is applied to the stator. As a result of this flux, the shading coil is induced with a voltage. This serves as a secondary winding. A coil's current direction should be such that it opposes the flux entering it, according to Lenz's law. In transformers, this acts as the secondary winding.

13. State the characteristics of shaded pole induction motor.

These are some of the characteristics of shaded pole motors.

- The starting torque is half that of a full-load machine
- Due to power loss in the shading coil, efficiency is low.
- Fans and other small devices use this motor
- The direction of rotation depends on the shaded coil position.

14. What are the advantages of shaded pole motor?

- Low cost,
- Self-starting
- Construction is simple
- Robust
- Reliability

15. What benefits do shaded pole motors offer?

Low cost, self-starting, straightforward in design, robust in nature, and reliability.

16. What are the disadvantages of shaded pole motor?

- Low starting torque
- Power factor is low
- More losses
- Poor efficiency
- Speed reversal is complicated

17. What drawbacks do shaded pole motors have?

Low power factor, extremely low starting torque, large losses, Lower efficiency and Difficult to reverse speed because it calls for pricey copper rings

18. What are the uses for an induction motor with shaded poles?

The following applications involve the usage of motors with shaded poles:

- They are perfect for small devices like relays and fans because they are inexpensive and simple to start.
- In table fans, hairdryers, and exhaust fans.
- In refrigeration, cooling fans, and air conditioning.
- Tape recorders, projectors, record players, electronic clocks, and photocopiers.

19. Why is the Shaded Pole Induction Motor suitable for low power rating?

This motor is not suitable for higher power ratings because of the following bottlenecks.

- There is a large magnetizing current drawn by the motor, resulting in a low power factor. Consequently, the motor has quite high no-load losses.
- Motors have very low starting torques.
- As a result of higher losses, the motor's efficiency is very low.

20. What will happen when we give an AC supply to DC series motor?

An AC supply will produce a unidirectional torque because the direction of both the currents (i.e. armature current and field current) reverses at the same time. Due to presence of alternating current, eddy currents are induced in the yoke and field cores which results in excessive heating of the yoke and field cores. Due to the high inductance of the field and the armature circuit, the power factor would become very low. There is sparking at the brushes of the DC series motor.

21. What are the advantages of hysteresis motor?

The main advantages of hysteresis motor are:

- No mechanical vibrations take place during its operation.
- Operate in quiet and noiseless.
- Suitable to accelerate inertia loads.
- Multi-speed operation is possible.

Short answer Questions with Answers:

1. **What is major difference between a 3-phase induction motor and a single-phase induction motor?**

   Unlike a 3-phase induction motor, a single-phase induction motor is not self- starting but requires some starting means.

2. **Why single phase induction motor does not self-starts?**

   The single-phase stator winding produces a magnetic field that pulsates in strength in a sinusoidal manner. The field polarity reverses after each half cycle but the field does not rotate. Consequently, the alternating flux cannot produce rotation in a stationary squirrel-cage rotor.

3. **What is the value of slip at standstill?**

   Note that each rotating field tends to drive the rotor in the direction in which the field rotates. Thus the point of zero slip for one field corresponds to 200% slip for the other. The value of 100% slip (standstill condition) is the same for both the fields.

4. **What is starting torque and starting current of split phase induction motor?**

   The starting torque is 15 to 2 times the full-loud torque and starting current is 6 to 8 times the full-load current.

5. **Why built-in-thermal relay is necessary in split phase induction motor and what is its purpose?**

   Since the starting winding is made of fine wire, the current density is high and the winding heats up quickly. If the starting period exceeds 5 seconds, the winding may burn out unless the motor is protected by built-in-thermal relay. This motor is, therefore, suitable where starting periods are not frequent.

6. **What is speed variation of split phase induction motor from no load to full load?**

   They are essentially constant-speed motors. The speed variation is 2-5% from no-load to full-load

7. **How do I identify a shaded pole motor?**

   A shading coil, which is a copper ring, can be used to identify this motor's auxiliary winding.

8. **Why is it called a shaded pole motor?**

A separate small winding, called a shading coil is used in the stator. The term shaded pole comes from that.

9. Is a shaded pole motor AC or DC?

Original type of AC single-phase induction motor.

10. What is the difference between shaded pole and PSC motors?

Shaded motors use old technology and old design, while PSC motors use new technology.

11. How do you control the speed of a shaded pole motor?

A dimmer can be used to control the speed of this motor by varying the voltage.

12. What are the applications of repulsion induction motor?

Repulsion induction motors are used in high starting torque with essentially a constant running speed.

13. What is a universal motor?

The motors which suit both a single-phase AC source and a DC source of supply and voltages are known as universal motors.

14. What are the advantages of the universal motor?

High power density is the main advantage of universal motors. The other benefits are low cost, portability, and capacity to function without control.

15. What are the types of a universal motor?

The two types of universal motors are compensated and uncompensated motors. The uncompensated is built with concentrated poles, and the compensated motor has field windings

16. Can I use a universal motor as a generator?

Universal motors can be used as DC series motors with modifications. But, universal motors are generally not recommended as generators due to voltage control issues. Modification required to use a universal motor as a DC generator is to add shunt winding to stabilize it.

17. What are the differences between DC motors and universal motors?

The universal motor has different winding ratios with thin iron laminations. DC motors cannot operate on AC supply, while universal motors operate on AC and DC supply.

18. What is hysteresis motor?

Hysteresis motor is defined as a synchronous motor that is having cylindrical rotor and works on hysteresis losses induced in the rotor of hardened steel with high retentivity. It is a single phase motor and its rotor is made of ferromagnetic material with nonmagnetic support over the shaft.

19. What is the speed-torque characteristics of hysteresis motor?

The torque is almost constant from starting to running condition. During starting the starting torque is the eddy current torque along with the hysteresis torque.

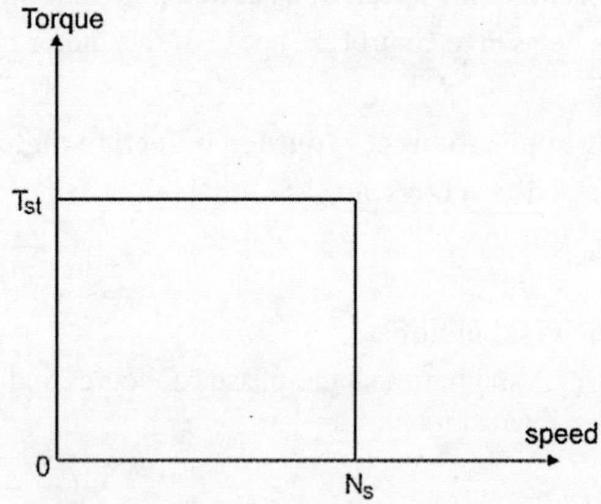

During running condition net running torque means only the hysteresis torque. The speed-torque characteristics is shown below.

Long answer Questions:

1. Discuss why single phase induction motor do not have a starting torque.
2. Draw the circuit diagram of a capacitor start capacitor run single phase induction motor and explain its working.
3. Using double revolving field theory, explain why a single phase induction motor is not self-starting.
4. What are the disadvantages of single phase induction motor when compared with a three phase induction motor?
5. Describe the construction and working of a shaded pole motor.
6. Explain the Construction of Repulsion Induction Motor.
7. Explain the Working of Repulsion Induction Motor.
8. What are the Characteristics of Repulsion Induction Motor?
9. What is an AC Series Motor?
10. What is conductively compensated type of motors?
11. Explain the working principle of the ac series motor.
12. Explain Principle of Operation of AC Series Motor.
13. Describe the constructional features of the ac series motor.

14. Draw the torque and speed characteristic of the ac series motor.
15. What are the Application of the AC Series Motor?
16. Explain briefly the maintenance of single phase induction motors.
17. What are the applications of hysteresis motor?
18. What are the disadvantages of hysteresis motor?

Short answer questions:

1. How to produce Rotating Magnetic Field from 2-Phase Supply?
2. Why split-phase induction motors are most popular single- phase motors in the market?
3. Is shaded pole motor self-starting?
4. What are the two types of AC series motor on the basis of the usage of compensating winding?
5. How does a universal motor work?
6. Explain the construction of hysteresis motor.
7. Explain the working principle of hysteresis motor.
8. What are the types of hysteresis motors?

Numerical Problems

1. A three-phase induction motor operates at 50 Hz with 6 poles. Determine the synchronous speed and slip when the motor is running at 1450 rpm.
2. In a resistance start induction run motor, the total resistance in the starting circuit is 8Ω, and the motor has 4 poles. Calculate the slip when the motor is running at 1100 rpm.
3. For a capacitor start capacitor run motor with 2 poles, the rated speed is 2800 rpm, and the full-load torque is 30Nm. Determine the slip at full load and the corresponding rotor speed.
4. A single-phase induction motor is operating at 230V and 60Hz, drawing a current of 6A with a power factor of 0.8. Calculate the apparent power, active power, and reactive power consumed by the motor.

PRACTICAL

1. Conduct experiments to measure and plot torque-speed characteristics for Resistance start induction run, capacitor start induction run, capacitor start capacitor run, shaded pole, repulsion type, series, universal, and hysteresis motors.
2. Demonstrate routine maintenance procedures for single-phase induction motors.
3. Emphasize tasks such as lubrication, bearing inspection, and cleaning.
4. Simulate common faults in single-phase induction motors and demonstrate the diagnostic process.
5. Discuss methods for rectifying identified faults.
6. Conduct an experiment on the given single phase induction motor to study the double field revolving theory.

KNOW MORE

1. Working of single phase Induction motor.
   https://www.youtube.com/watch?v=awrUxv7B-a8
2. Working of shaded pole induction motor.
   https://www.youtube.com/watch?v=4EYwvV9yM30
   https://www.youtube.com/watch?v=_vy4zGs0ynw
3. Working of repulsion motor.
   https://www.youtube.com/watch?v=0Clwa6SGesk
4. Working of AC series motor
   https://www.youtube.com/watch?v=8QUCqG3O8gA
5. Working of universal motor
   https://www.youtube.com/watch?v=1T_SQIO-1Xg

REFERENCES AND SUGGESTED READINGS

1. P.S. Bimbhra, Electric Machines, Khanna Book Publishing Co., New Delhi (ISBN: 978-93- 6173- 294)
2. Mittle, V.N. and Mittle, Arvind, Basic Electrical Engineering, McGraw Hill Education New Delhi, ISBN :9780070593572
3. Kothari, D. P. and Nagrath, I. J., Electrical Machines, McGraw Hill Education. New Delhi, ISBN:9780070699670
4. Bhattacharya, S. K., Electrical Machines, McGraw Hill Education, New Delhi, ISBN:9789332902855
5. Theraja, B.L., Electrical Technology Vol-II (AC and DC machines), S.Chand and Co. Ltd., New Delhi, ISBN : 9788121924375
6. Sen, S. K., Special Purpose Electrical Machines, Khanna Publishers, New Delhi, ISBN: 9788174091529
7. Janardanan E. G, Special Electrical Machines, Prentice Hall India, New Delhi ISBN: 9788120348806
8. Hughes E., Electrical Technology, ELBS
9. Cotton H., Electrical Technology, ELBS

Dynamic QR Code for Further Reading

5 THREE PHASE ALTERNATOR

Working and Construction

UNIT SPECIFICS

Through this module we have discussed the following aspects:

- *Constructional details of 3-phase alternator;*
- *Concept of cylindrical type rotor and salient pole type rotor and their speed;*
- *Familiarizing the working of alternator;*
- *EMF equation of alternator;*
- *Coil span factor and distribution factor;*
- *Problems.*

The figures of alternator are shown to provide the details of stator stampings, and various parts of the rotor. Better understanding on EMF generation through numerical problems. A list of references, animated videos and suggested readings are given in the unit so that one can go through them for practice.

RATIONALE

Alternator is the Electric power provider without which the global survival has become a question today. It has a widespread application. This chapter introduces the students to the construction, working principle and generation of EMF of the alternator.

PRE-REQUISITES

Machines I:Sem III

Physics: Electromagnetism (Class X)

UNIT OUTCOMES

List of outcomes of this unit is as follows:

At the end of the chapter student will be able to

U5-O1: explain the constructional difference in smooth cylindrical rotor and salient pole rotor.

U5-O2: understand the working of an alternator

U5-O3: describe coil span factor and distribution factor.

U5-O4: differentiate short pitched coil and full pitched coil.

U5-O5: derive an expression for EMF.

Unit-5 Outcomes	***EXPECTEDMAPPINGWITHCOURSEOUTCOMES*** *(1-WeakCorrelation;2-Mediumcorrelation;3-StrongCorrelation)*					
	CO-1	***CO-2***	***CO-3***	***CO-4***	***CO-5***	***CO-6***
U5-O1	3	-	-	-	-	-
U5-O2	*3*	-	-	-	-	-
U5-O3	*3*	-	-	-	-	-
U5-O4	*3*	-	-	-	-	-
U5-O5	*3*	-	-	-	-	-

5.1 INTRODUCTION

An AC generator is a machine that transforms mechanical energy from a prime mover into Alternating Current (AC) electrical power. Synchronous generators, which operate as three-phase AC machines play a pivotal role in producing significant power quantities in thermal, hydroelectric, and nuclear power plants. They run at a constant speed called synchronous speed. Two distinct categories of alternators exist: Revolving armature type and the revolving field type. In the case of revolving armature alternators, the field poles remain fixed while the armature serves as the rotor. This configuration is suitable for applications with lower capacity and voltage ratings. Its design resembles that of a DC generator, employing slip rings instead of split rings (commutator). In revolving field type, the armature forms the stator and the field poles rotate. Most alternators belong to this classification. The advantages of this type of stationary armature are as follows:

- The armature windings can be protected against mechanical stresses due to centrifugal forces and short circuit currents.
- Insulating the armature winding for high voltage is imperative and it is easier to insulate the high voltage armature winding, as it is stationary.
- The armature is cooled more easily and readily.

5.2 CONSTRUCTIONAL DETAILS OF ALTERNATOR

The main components of alternator are as follows:

a. Stator(stationary):
   1. Stator core
   2. Stator winding
   3. Stator frame

b. Rotor (rotating):
   1. Magnetic poles and field winding
   2. Slip-rings
   3. Brush and brush holder
   4. Spider
   5. Exciter
   6. Shaft and bearings

a. Stator

The stator core, constituting the immobile element of the alternator, remains stationary. It is constructed from the laminations of silicon steel (a steel alloy) as given in Figure 5.1. The stator core is the armature core. Along the innermost perimeter of these laminated stator cores, the slots are punched. These slots serve as the house for assembling the stator winding (also known as the armature winding). The entire assembly, comprising the stator core and winding, is upheld by a stator frame.

The alternator's rotating magnetic field continuously interacts with the stator core. This interaction generates circulating electric currents called eddy currents, which result in power losses within the core. The stator core is thus constructed using thin laminated stampings(sheets) insulated from one another to prevent the eddy currents from forming and causing losses. For smaller machines, these stampings are cut into complete rings, and for larger machines, they are divided into segments. The sheets also have holes for ventilation, both along the length and from the center outward, to help with effective cooling.

On the inside rim of the stator core, there are slots available. There could be open type slots or semi-closed slots. Due to the possibility of form-wound coils, the majority of alternators use open type slots. In case of conductor failure, burnt coils can be easily replaced in this open type slots. The winding approach is more effective. However, the distribution of air-gap flux will not be uniform, leading to distortion in the produced EMF waveform. Semi-closed slots can be used to get around this, however they cannot accommodate the use of form-wound coils. Fully closed slots, although effective, are more expensive to implement and are thus less commonly used.

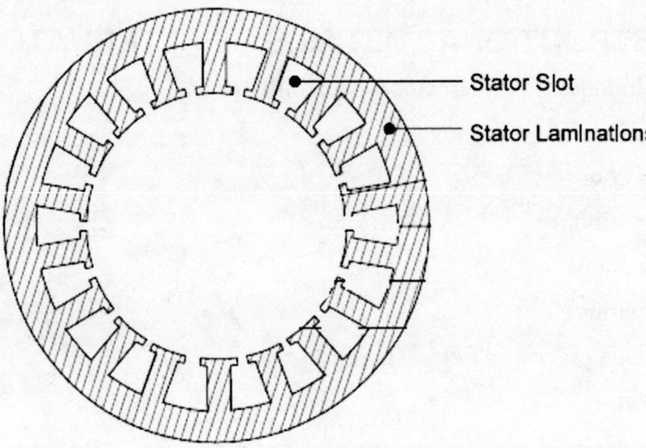

Fig.5.1: Stator stamping

Insulated copper or aluminium conductors are inserted into the slots in the stator core to assemble the armature winding. When the magnetic field cuts these stationary conductors, they induce EMF. An alternator's armature basically consists of three-phase winding in a delta or star fashion for connection.

The outermost component of the alternator is called the stator frame. This frame plays a crucial role in providing support to the entire alternator structure. It is usually made from welded steel plates, cast iron, or cast steel materials. One of its key functions is to prevent the dust and other particles from entering the sensitive parts of the alternator. Moreover, the stator frame helps to maintain the proper positioning of the stator core and winding components. In addition to this, it offers mechanical protection to the entire alternator system.

b. Rotor

The rotating part of the alternator is called the rotor. It has copper windings, also referred to as field winding, and is manufactured in the form of a cylinder. The spinning magnetic field is produced by the field windings, which are electromagnets. The shaft of the rotor is turned by a driving belt and pulley arrangement. A prime mover is the source that turns the rotor. Any type of device, including an engine, a water or wind turbine, is possible. In alternators or synchronous generators, there are two different types of rotors. In alternators or synchronous generators, there are two main types of rotors: Salient Pole and Cylindrical Pole. These types of rotors have distinct designs and functions.

Salient Pole Type:

This rotor style features a significant number of projection or projecting poles installed on a magnetically laminated steel or cast-iron core. The word "salient" describes something that protrudes or is projected, like in the illustration given in Figure 5.2 below. To minimize the eddy current losses, these poles are made from laminated steel or cast iron with strong magnetic properties.

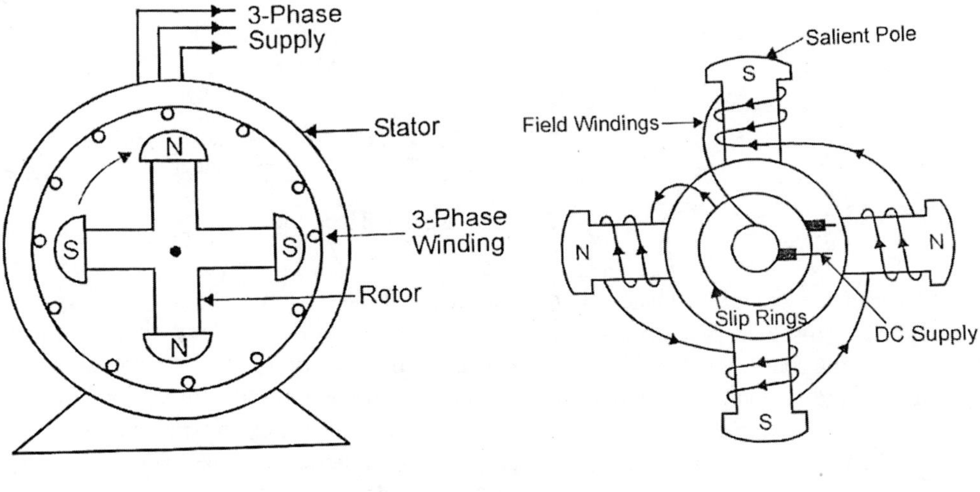

Fig.5.2 – Salient pole rotor

The pole shoes include numerous slots for winding dampers, which help to prevent haunting. After being wound across the poles, the field coils are joined in series, the ends of the coils are connected to a separate DC power source through a set of slip rings, the field coil is powered. On the rotor's shaft are mounted the slip ring and brushes. Salient pole type alternators are characterized by short axial length and larger diameter. Due to this, they are employed as low and medium speed alternators, like those found in hydroelectric plants. The pole face is shaped in a way that the gap between the pole center and pole tips increases radially. This arrangement ensures an even distribution of magnetic flux across the armature, resulting in the creation of a smooth sinusoidal electrical waveform. Additionally, the pole shoe takes up about two-thirds of the pole pitch.

Smooth Cylindrical Type:

This kind of rotor has two or four poles. These alternators are commonly linked to high-speed steam turbines. The rotor is built using a steel cylinder layered with laminations. On the outer edge of the rotor, there are slots punched to hold the field winding. Around two-thirds of the rotor's pole pitch has these slots, while the remaining one-third doesn't. The field windings on this cylindrical rotor are connected in series fashion.

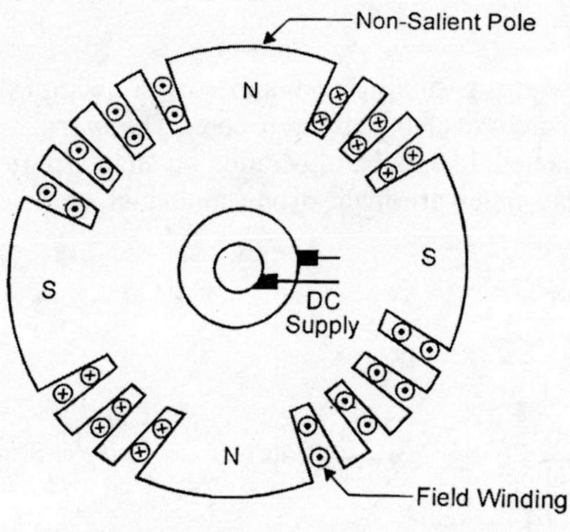

Fig.5.3: Cylindrical type rotor

As seen in Figure 5.3 illustration, the pole areas of the rotor are not slotted. This type is referred to as a non-salient pole or round rotor because the poles don't extend out from the core. Unlike the salient pole rotor, this kind possesses fewer protruding poles, contributing to a greater axial length and a smaller rotor diameter. Given its lengthier rotor core, natural ventilation is not enough, and external cooling methods are essential for optimal performance. Robust construction, mechanical strength, resilience, noise free operation, better dynamic balancing, and a consistent distribution of magnetic flux are all provided by the smooth cylindrical design. Less windage loss occurs. It is appropriate for operations that require high speed (around 3000 rpm) and silence. Damper windings are unnecessary in this design. These cylindrical designs are well-suited for high-speed alternators, commonly found in thermal power plants.

5.3 OPERATION OF SYNCHRONOUS GENERATOR

Generation of alternating voltage

An alternating voltage gets produced when an armature coil rotates within a steady magnetic field created by unmoving field poles. This type of voltage can also be generated when a stationary armature is surrounded by magnetic field poles that are in motion. Whenever the armature conductors move in relation to the field's magnetic flow, it creates voltage in those conductors. In both situations, the resulting voltage takes on a smooth wave pattern, resembling the natural curves of a sine wave.

In direct current (DC) generators, the field poles remain still while the armature conductors spin. The produced voltage, even though it's alternating, is transformed into a steady direct voltage at the brushes by utilizing commutator. Unlike a DC generator, rotating magnetic field and a stationary armature coil make up the Synchronous Generator or alternator. It operates on a basic principle of electromagnetic induction as DC generators.

The alternator's rotor is propelled at the appropriate speed by the prime mover, a device that provides the necessary mechanical energy to it. In the case of slower and moderately paced alternators, water turbines are commonly employed as the prime mover, while steam and gas turbines are preferred for larger and faster alternators. As the rotor spins, its poles move beneath the armature conductors of the stationary stator. Consequently, the unchanging armature conductors are cut by the magnetic flux. This interaction induces an Electro Motive Force (EMF) and thereby circulates the current within them. Due to the alternation between north pole and south magnetic poles, one full cycle of alternating EMF is produced in the armature coil when a north pole and a south pole cross the coil. This induced EMF within the synchronous generator or alternator can be calculated using the equation derived as presented below.

Relationship between Number of poles, Frequency and Speed

The frequency of the generated EMF depends on two factors: the count of the poles in the field system and the speed of the field poles' movement. During one complete rotation of the field system, an armature coil experiences cut by P/2 north poles and P/2 south poles. Whenever a pair of poles goes over a conductor, it induces one cycle of EMF in it. This corresponds to an angular distance that is twice the pole pitch. The number of cycles generated per revolution of field pole is number of field pole pairs, which is P/2.

Frequency (f) = (Number of cycles generated per Revolution) x (Number of Revolutions made per second).

$$f = (P/2) \text{ x } (N/60) = PN/120\text{Hz}.$$

Where,

P = Total number of magnetic poles

N = Rotor speed in RPM

$N = 120f/P$

f = Frequency of the generated EMF in Hz

To achieve the desired frequency in its output, the alternator needs to operate at its synchronous speed. This requirement stems from the fact that the positions of the rotor poles remain fixed.

Expression for induced EMF in a synchronous machine

During one revolution of the rotor, each stator conductor cuts a flux of $d\varphi = P\varphi$ webers.

The time it takes for the rotor poles to complete one full revolution is $dt = \frac{60}{N}$ seconds. Hence, in accordance with Faraday's law of electromagnetic induction, the average electromotive force (EMF) induced in a single stator conductor is given by

$$\text{Average EMF per conductor} = \frac{d\varphi}{\text{dt}} = \frac{P\varphi}{\left(\frac{60}{N}\right)} = \frac{P\varphi\text{N}}{60} \text{weber/sec}$$

There are 'Z' conductors in series per phase. Average induced EMF per phase,

E_{av} = EMF per conductor × Number of conductors

$$E_{av} = \frac{P\varphi\text{N}}{60} Z \text{ volts}$$

We know that the rotor speed, $N = \frac{120f}{P}$

$$\therefore E_{av} = \frac{P\varphi\left(\frac{120f}{P}\right)}{60} Z = 2\varphi f Z \quad \text{-------(1)}$$

Equation (1) provides the average value of induced electromotive force (EMF) per phase. Considering that each turn consists of two conductors (Z = 2T), the formula for the average induced EMF per phase can be expressed as:

$E_{av} = 4T\varphi f$ volts

Where,

Z = Number of conductors in series per phase

T = Number of coils or turns per phase

φ = Flux per pole in webers

RMS value of EMF per phase = (Average Value/Phase) × (Form Factor)

For a sinusoidal voltage wave, form factor = 1.11.

The RMS value of the induced EMF per phase can be written as,

$$E_{rms} = 4T\varphi f x\ 1.11 = 4.44\varphi f T \text{ volts} \quad \text{-------(2)}$$

Equation (2) gives the RMS value of the induced EMF with the following assumptions –

- All the armature conductors are concentrated in one stator slot.

- Windings per phase is full pitched.

The winding being short pitched and distributed, the actual induced EMF available per phase takes the coil span factor and distribution factor into account.

The actual induced EMF per phase is thus given by,

$$E_{ph} = 4.44\varphi f T k_d k_p \text{ volts} \quad \text{-------(3)}$$

Where,

k_d = Distribution factor
k_p = Pitch factor

Equation (3) is called as the EMF equation of an alternator.
In a star-connected alternator:

- The line voltage is √3 times the phase voltage: $E_L = \sqrt{3}E_{ph}$.
- However, the line current and phase current remain the same: $I_L = I_{ph}$.

In a delta-connected alternator:

- The line current is √3 times the phase current: $I_L = \sqrt{3}I_{ph}$.
- Yet, the line voltage and phase voltage are equivalent: $E_L = E_{ph}$.

Armature Windings

The coil that carries current to create the main magnetic field is referred to as the field winding. On the other hand, the coil in which an electromotive force (EMF) is generated is known as the armature winding. This is a contrast to direct current (DC) machines, which employ closed circuit windings. In alternators, the windings are open, meaning that there is no continuous closed path for the armature currents within the winding itself. One end of the winding is connected to the neutral point, while the other end is brought out, especially for a star-connected armature. There are two types of windings: single layer and double layer. The single layer windings come in two forms: wave winding or chain winding. Below, you'll find explanations of some fundamental terms related to windings.

- **Pole pitch:** This is the angular distance between the centers of two neighboring poles in a machine.
- **Coil span:** It's the distance, measured in terms of either armature slots or conductors, between the two sides of a coil.
- **Full pitch winding:** When a coil has a span equal to one pole pitch (covering 180° electrically), it's referred to as a full pitch winding. For instance, if the coil sides are placed in slots 1 and 7, it's a full-pitched winding. If the coil sides are in slots 1 and 6, then it's short-pitched. This means the coil span is 5/6 of the pole pitch or 30° less than 180°. It falls short by 1/6. This is illustrated in the Figure 5.4.

- **Short pitch winding:** Coils are considered short-pitched when their span is slightly less than the pole pitch, meaning it is less than 180°. In this case, the coils are also called chorded or fractionally pitched. This is illustrated in the Figure 5.4.

Advantages of short-pitched coils

- They reduce the amount of copper needed for end connections.
- The generated EMF can be made to resemble a sine wave more closely, reducing or eliminating the distorting harmonics.
- Efficiency is improved due to a reduction in eddy current and hysteresis losses caused by the elimination of high-frequency harmonics.

Disadvantages of short-pitched coils

The overall voltage around the coil is somewhat decreased because the voltages induced in the two sides of short-pitched coils are slightly out of phase. This results in the vectorial sum being less than their arithmetic sum.

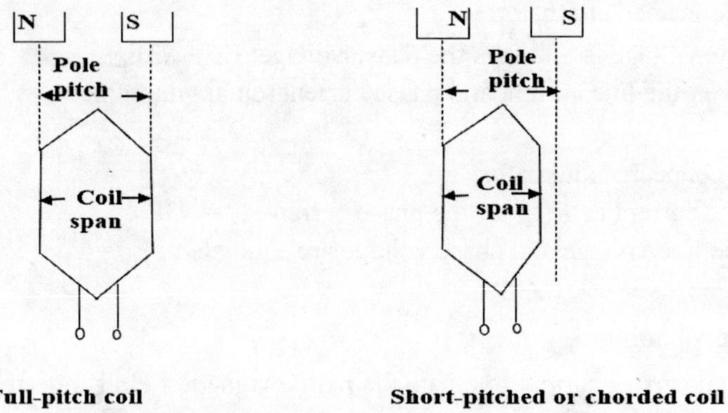

Fig.5.4: Full pitched and Short Pitched Coil

Turns and Conductors

A turn consists of two conductors. A coil is made of several turns connected in series as given in Figure 5.5.

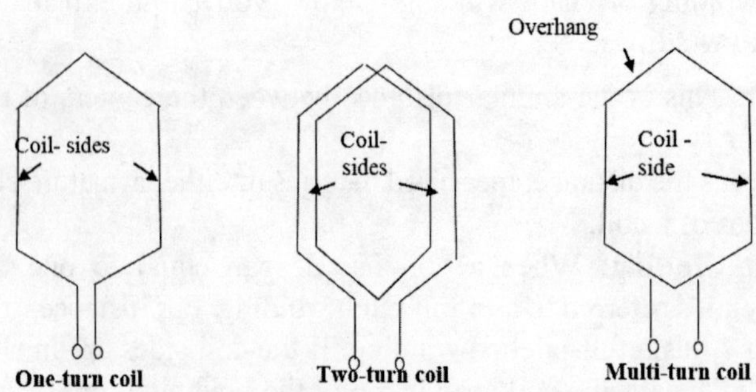

Fig.5.5: Relationship between Number of Turns and Conductors

$$\text{Coil span factor} = \frac{\text{Vector sum of the EMF induced per coil}}{\text{Arithmetic sum of the EMF induced per coil}}$$

Coil span factor or pitch factor for short pitched windings (k_p)

The total EMF induced per phase is lesser in short pitch windings coils compared to that of full pitch winding coils. The reduction in EMF is quantified using pitch factor,

$$\text{Pitch factor,} \quad k_p = \frac{\text{EMF induced in a short pitch coil}}{\text{EMF induced in a full pitch coil}}$$

If the coil span falls short of full pitch by an angleα,

$$k_p = \cos\frac{\alpha}{2}$$

$$k_p = \cos\frac{\alpha}{2} = \frac{180 * Number\ of\ slots\ for\ which\ the\ winding\ is\ short\ pitched}{\left(\frac{Number\ of\ slots}{Number\ of\ poles}\right)}$$

Where,

k_p, Pitch factor is lesser than unity.

$\alpha$is short pitch angle.

Distribution factor or breadth factor or spread factor (k_d)

When the coils comprising a phase of windings are distributed in two or more slots per pole, the EMF in the adjacent coils is out of phase with respect to one another and their resultant will be lesser than the algebraic sum.

$$\text{Distribution factor,} k_d = \frac{\text{EMF with distributed winding}}{\text{EMF with concentrated winding}}$$

$$k_d = \frac{Vector\ sum\ of\ EMF\ induced\ per\ coil}{\text{Arithmetic sum of EMF induced per coil}} = \frac{\sin\frac{m\beta}{2}}{m\sin\frac{\beta}{2}}$$

Where,

$k_d$is lesser than unity for distributed winding.

m = number of slots per pole per phase.

β = angular displacements between the slots.

$$\beta = \frac{180}{\left(\frac{Number\ of\ slots}{Number\ of\ poles}\right)}$$

UNIT SUMMARY

The aim of this chapter is to understand the construction and working principle of the alternator. The alternator has two types of construction which is deployed based on the requirement of the application, speed, and the prime mover. The principle of operation of the alternator conveys clearly why alternator is called as a synchronous machine. Also, discussed about the EMF equation and numerical in the alternator.

EXERCISES

1. Determine the speed at which a 6-pole alternator should be driven to obtain the frequency of EMF induced to be 60Hz.

Given Data:

Number of poles (P) = 6, frequency (f) = 60Hz.

Solution:

$N = 120\, f/P$=120 x 60 / 6 = 1200rpm.

2. Calculate the mechanical and electrical degree between adjacent poles in a 6-pole electrical machine.

Given Data:

Number of poles (P) = 6

Solution:

Mechanical degree or angular measure in space (Θ_{mech}) = 360/P = 360/6 = 60°.

Electrical degree or angular measure in cycles (Θ_{elec}) = (P/2) x Θ_{mech}= 6/2 x 60° = 180°.

3. A 3 phase, 16 pole, star-connected alternator has 240 stator slots with 8 conductors per slot and the conductor of each phase are connected in series. The coil span is 144° electrical. The machine speed is at 375rpm and the flux per pole is 0.061wb distributed in the air gap sinusoidally. Calculate the phase and line EMFs.

Given Data:

Number of poles (P) = 16, Y-connected, Number of slots (S) = 240, Number of conductors per slot

(Z_S) = 8, Speed (N) = 375rpm, flux (φ) = 0.061 wb.

Solution:

$\frac{Number\ of\ slots}{Number\ of\ poles}$ = S/P = 240/16 = 15.

Number of slots per pole perphase (m) = 15/3 = 5

Angular displacements between the slots $(\beta) = \frac{180°}{(\frac{Number\ of\ slots}{Number\ of\ poles})}$ = 180°/15 = 12°electricalDistribution factor, $k_d = \frac{\sin\frac{m\beta}{2}}{m\sin\frac{\beta}{2}}$

$$k_d = \frac{\sin\frac{5\,x\,12°}{2}}{5\sin\frac{12°}{2}} = 0.957$$

Pitch factor, $k_p = \cos\frac{\alpha}{2}$

Chording angle, $\alpha = 180° - \text{coilspan} = 180° - 144° = 36°$

$$k_p = \cos\frac{36°}{2} = 0.951$$

Frequency, $f = \frac{PN}{120} = \frac{16\,x\,375}{120} = 50\text{Hz}$

Number of turns per phase (T) $= \frac{Number\ of\ conductors\ per\ slot\ x\ Number\ of\ slots}{2\ x\ Number\ of\ phases}$

$$T = \frac{Zs \times S}{2 \times 3} = \frac{8 \times 240}{2 \times 3} = 320$$

Induced EMF per phase(E_{ph})$= 4.44\varphi f T k_d k_p$ volts

$E_{ph} = 4.44 \times 0.061 \times 50 \times 320 \times 0.957 \times 0.951 = 3943.43$ volts

Line voltage (E_L) $= \sqrt{3}E_{ph} = 6830.02$ volts.

Multiple Choice Questions

1. Alternator operates on principles of
   a. Electromagnetic induction
   b. Self-induction
   c. Mutual induction
   d. Self or mutual induction

2. In alternators, the rotating part is
   a. Armature
   b. Field system
   c. Armature as well as field system
   d. None of the above

3. Salient pole type rotors are
   a. Larger in diameter and larger in axial length
   b. Larger in diameter and smaller in axial length
   c. Smaller in diameter and larger in axial length
   d. Smaller in diameter and smaller in axial length

4. Distributed winding is preferred over concentrated winding as it
   a. Reduces noise
   b. Reduces amount of copper required
   c. Reduces harmonics thus improves generated emf waveform

d. Reduces machine size

5. The emf generated in alternator due to nth harmonic components of flux will be ________ the fundamental emf in magnitude.
   a. Equal to
   b. More than
   c. Less than
   d. None of the above

Answers of Multiple Choice Questions

1. a. Electromagnetic induction
2. b. Field system
3. b. Larger in diameter and smaller in axial length
4. c. Reduces harmonics thus improves generated emf waveform
5. c. Less than

Short and Long Answer Type Questions

Long Answer Questions with Answers

1. **What are the advantages of stationary armature and rotating field system? Show the difference between stationary armature and rotating armature types.**

Advantages of stationary armature and rotating field

- Ease of construction.
- Better insulation of armature possible.
- Lesser no of slip rings-Only two slip rings are required for DC supply to the field winding on the rotor.
- Thicker conductors can be used in the armature.
- Reduced rotor inertia.
- Increased output.
- Easy of current collection.
- Reduced armature leakage reactance.
- Due to simple and robust construction of the rotor, higher speed of rotating DC field is possible.
- Improved ventilation and heat dissipation.

Advantages of stationary armature are as follows

- The armature windings can be protected against mechanical stresses due to centrifugal forces and short circuit currents.
- Insulating the armature winding for high voltage is imperative and it is easier to insulate the high voltage armature winding, as it is stationary.

The armature is cooled more easily and readily.

The difference between stationary armature and rotating armature types are given in the figures below

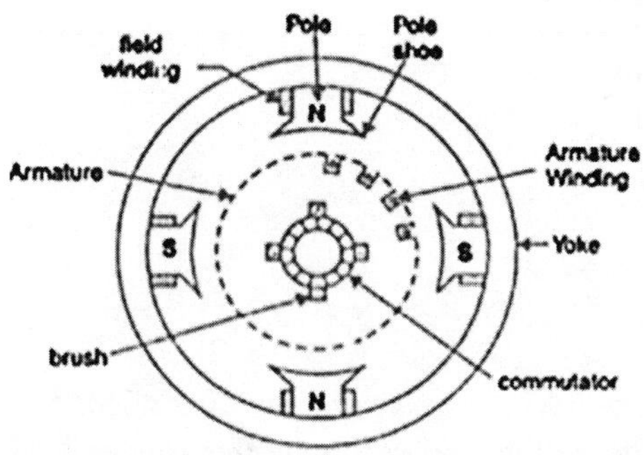

Fig 5.6 DC Generator: Rotating Armature and Stationary Field

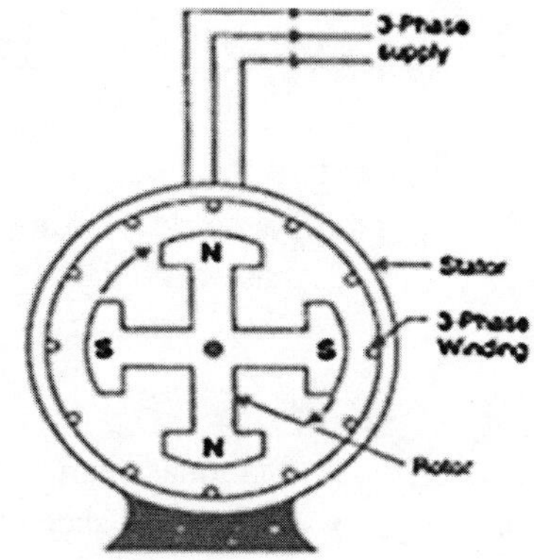

Fig 5.7 AC Generator: Stationary Armature and Rotating field

2. Differentiate between cylindrical pole and salient pole rotor.

Cylindrical pole rotor	Salient pole rotor
• Rotor surface is smooth	• Due to projected poles, rotor surface not smooth
• Used for high-speed turbo generators such as the ones driven by steam turbines or gas turbines	• Used for low or medium speed hydro-generators driven by water-turbines
• Most common speed range is 3000 or 1500 rpm	• Most common speed range is 100 – 400 rpm
• Flux distribution in space between the rotor and the stator is sinusoidal as the air gap is uniform between stator and rotor	• Unless the rotor is specially shaped, the flux distribution between stator and rotor may not be sinusoidal due be non-uniform air gap
• Since the speed is high, to generate rated frequency, the number poles in rotor are restricted to 2 – 4	• Since the speed is low, to generate rated frequency, it is necessary to use large number of poles in rotor, starting from 4 to 50
• To restrict centrifugal forces at high speed, diameter of the rotor is lower	• To accommodate large number of poles, diameter of rotor is large
• Since the diameter is low, to produce the desired output, axial length of rotor is made large as compared to diameter	• Since diameter is high, axial length rotor is kept lower to produce the desired output

• Friction of the rotating rotor with air is low	• Air friction while rotation is much higher
• Rotation is less noisy	• Rotation produces lot of noise

3. Explain the construction of alternator

Winding construction

- 3-phase winding on the stator
- DC field winding on the rotor

Stator:

- Stationary part of the machine.
- Built up of sheet-steel laminations having slots on its inner periphery.
- 3-phase winding is placed in these slots and serves as the armature winding of the alternator.
- Armature winding is connected in star and the neutral is connected to ground

Rotor:

- Rotor carries a field winding which is supplied with DC source through two slip rings.
- DC source (called exciter) is generally a small DC shunt or compound generator coupled with shaft of the alternator.

Types of rotor

i) Salient (or projecting) pole type and ii) Non-salient (or cylindrical) pole type. The constructional layout of both types are given in figures below.

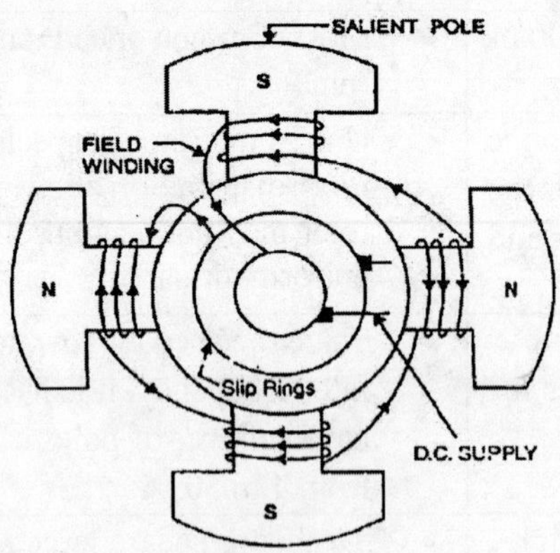

Fig 5.8 Salient Pole type rotor

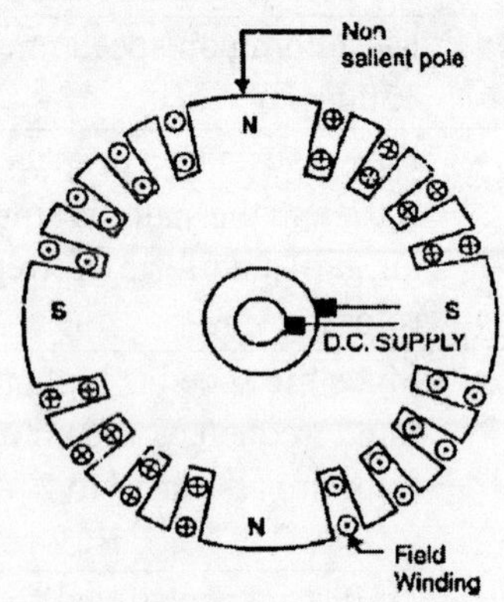

Fig 5.9 Non salient pole type rotor

i) Salient pole type rotor:

- Salient or projecting poles are mounted on a large circular steel frame.
- Fixed to shaft of alternator.
- Individual field pole windings are connected in series.

- When field winding is energized by the DC exciter, adjacent poles have opposite polarities.
- Low and medium-speed alternators (120-400 rpm) such as those driven by diesel engines or water turbines have salient pole type rotors due to the following reasons:
  a) Excessive wind age loss if driven at high speeds and would tend to produce noise.
  b) Salient-pole construction cannot withstand the mechanical stresses when subjected to higher speeds.
  c) Salient-pole type rotors have large diameters and short axial lengths.

ii) Non-salient pole type rotor

- Rotor is made of smooth solid forged-steel radial cylinder.
- Has a number of slots along the outer periphery.
- Field windings are embedded in these slots and are connected in series to the slip rings.
- Field windings are energized by the DC exciter.
- The regions forming the poles are usually left unslotted.
- Poles formed are non-salient.
- High-speed alternators (1500 or 3000 rpm) are driven by **steam turbines** and use non-salient type rotors due to the following reasons:
  a) Mechanical robustness and gives noiseless operation at high speeds.
  b) The flux distribution around the periphery is nearly a sine wave
  c) Better emf waveform is obtained compared to salient-pole type.
  d) Steam turbines run at high speed and a frequency of 50 Hz is required.
  e) So, small number of poles on the rotor are used. (also called turbo alternators).
  f) Has 2 or 4 poles and have small diameters and very long axial lengths.

Numerical Problems

1. Explain the effect of harmonics on pitch and distribution factors.
2. Find the pitch factor for the alternator with four poles, coil span 1 to 8, and 36 stator slots.
   [**Ans:** 0.94]
3. Determine distribution factor for the alternator having 4-pole, 36 slots, three phase single layer winding.
   [**Ans**:0.96]
4. An alternator has 10 pole, 50Hz, 600 rpm, armature diameter 1.2m, core length 0.4m, 180 slots wound with double layer, coil span of 15 slots, three turn. The coils are connected in 60deg. Flux density distribution is a function of $B = \sin\Theta + 0.5\sin3\Theta + 0.3\sin5\Theta$. Derive the equation for i) instantaneous emf per conductor, ii) instantaneous emf per coil.
   [**Ans:** i) $e = 14.93\sin\Theta + 5.97\sin 3\Theta + 2.98\sin5\Theta$. ii) $86.5\sin\Theta + 25.3\sin3\Theta + 4.63\sin5\Theta$]
5. A 6pole alternator has 90 slots, 8 conductors per slot, flux per pole 50mwb revolves at 1000rpm. Find the generated EMF assuming star connection machine with winding factor 0.96.
   [**Ans:** 1280V]

PRACTICAL

1. Showcase the application of moving armature principles in small DC motors used in household appliances or toys. Disassemble and analyze a small DC motor to identify its components.
2. Compare the rotor constructions of different types of motors (e.g., squirrel cage rotor of induction motor vs. wound rotor of synchronous motor). Discuss the advantages and applications of each.
3. Discuss the practical applications of single and double-layer windings in motors and generators. Compare the winding configurations in terms of efficiency and performance.
4. Use a small alternator model to explain the key components, including the rotor, stator, and winding arrangement. Measure the voltage output at different speeds to understand the relationship.

KNOW MORE

1. **How an alternator works**
   https://youtu.be/QXlOSRjJT8g?si=fgAmOGf41tvdJdEx
   https://youtu.be/T8zJKXkU_yM?si=AbcCCfunS0pdNViY
2. **Construction of synchronous machine**
   https://youtu.be/QBpN09tWv2U?si=rYQn6TQi51EOWcqA

REFERENCES AND SUGGESTED READINGS

1. P.S. Bimbhra, Electric Machines, Khanna Book Publishing Co., New Delhi (ISBN: 978- 93-6173- 294)
2. Mittle, V.N. and Mittle, Arvind., Basic Electrical Engineering, McGraw Hill Education New Delhi, ISBN :9780070593572
3. Kothari, D. P. and Nagrath, I. J., Electrical Machines, McGraw Hill Education. New Delhi, ISBN:9780070699670
4. Bhattacharya, S. K., Electrical Machines, McGraw Hill Education, New Delhi, ISBN:9789332902855
5. Theraja, B.L., Electrical Technology Vol-II (AC and DC machines), S.Chand and Co. Ltd., New Delhi, ISBN : 9788121924375
6. Hughes E., Electrical Technology, ELBS
7. Cotton H., Electrical Technology, ELBS

Dynamic QR Code for Further Reading

6 THREE PHASE ALTERNATOR- Voltage Regulation

UNIT SPECIFICS

Through this module we have discussed the following aspects:

- *Alternator loading-Factors affecting the terminal voltage of alternator;*
- *Armature resistance, and leakage reactance drops;*
- *Voltage Drop due to Armature reaction;*
- *Armature reaction at various power factors;*
- *Terminal Voltage Equation - Synchronous impedance;*
- *Phasor diagram of Alternator under various load conditions;*
- *Voltage regulation- synchronous impedance method;*
- *Maintenance of alternators.*

The figures of alternator are shown to provide the Equivalent circuit of alternator, Details of the effect of load power factor on armature reaction, Phasor diagram of Alternator under various load condition, Open circuit and short circuit characteristics of the alternator.

RATIONALE

Alternator is the Electric power provider without which the global survival has become a question Today. It has a widespread applications. This chapter details to the students about the alternator on load, terminal voltage equation of the alternator under various load conditions construction, and voltage regulation of the alternator.

PRE-REQUISITES

Machines I:Sem III

Physics: Electromagnetism (Class X)

UNIT OUTCOMES

List of outcomes of this unit is as follows:

At the end of the chapter student will be able to

U6-O1: Explain the factors affecting the terminal voltage of alternator.

U6-O2: Understand the armature reaction effect on main flux.

U6-O3: Derive terminal voltage equation for alternator under load.

U6-O4: Determine voltage regulation of alternator.

U6-O5: Understand the basics of alternator maintenance.

Unit-6 Outcomes	EXPECTEDMAPPINGWITHCOURSEOUTCOMES (1-WeakCorrelation;2-Mediumcorrelation;3-StrongCorrelation)				
	CO-1	CO-2	CO-3	CO-4	CO-5
U6-O1	-	-	-	3	-
U6-O2	-	-	-	3	-
U6-O3	-	-	-	3	-
U6-O4	-	-	-	3	-
U6-O5	-	-	-	3	-

6.1 INTRODUCTION

Alternators are crucial components in electrical power generation. The performance of these is subject to countless factors that influence their terminal voltage. Understanding and managing these factors is essential for ensuring a stable and reliable power supply. This introduction delves into the details of alternator loading, exploring key elements such as armature resistance, leakage reactance drops, armature reaction at different power factors, synchronous impedance, and voltage regulation methods. The discussion extends to direct loading and synchronous impedance approaches for maintaining optimal voltage levels. Lastly, the importance of regular maintenance practices in sustaining the efficiency and longevity of alternators is highlighted. Together, these aspects form a comprehensive overview of the factors that shape the terminal voltage of alternators, shedding light on the particulars involved in their operation and maintenance.

ALTERNATOR LOADING

6.1.1 Factors affecting the terminal voltage of alternator

When an alternator is put under load, the entire induced electromotive force (EMF), ie., **Eg** doesn't fully appear across the terminals of the output, $\mathbf{V_t}$. As the load on an alternator change, the terminal voltage of the alternator also changes, and it is evident from the equivalent circuit of the alternator in Figure 6.1. This variation in terminal voltage is caused by different factors, including voltage drops due to the armature's resistance (Ra), armature leakage reactance (X_L), and the influence of the armature reaction.

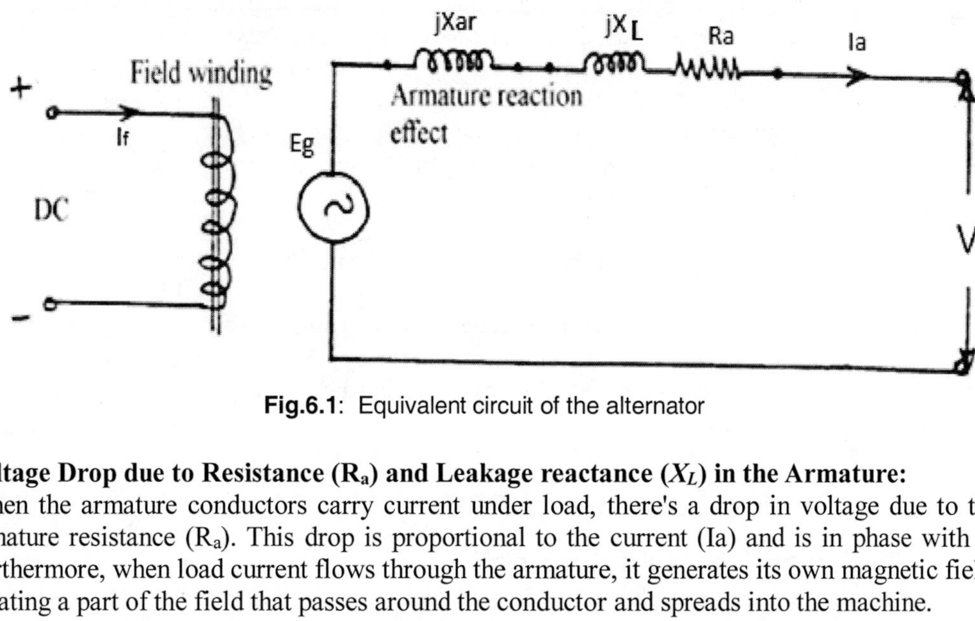

Fig.6.1: Equivalent circuit of the alternator

Voltage Drop due to Resistance (R_a) and Leakage reactance (X_L) in the Armature:
When the armature conductors carry current under load, there's a drop in voltage due to the armature resistance (R_a). This drop is proportional to the current (Ia) and is in phase with it. Furthermore, when load current flows through the armature, it generates its own magnetic field, creating a part of the field that passes around the conductor and spreads into the machine.

These additional magnetic fields, known as leakage fluxes, include various types like slot leakage flux, zig-zag leakage flux, harmonic leakage flux, peripheral leakage flux, tooth top leakage flux, and

skew leakage flux. These leakage fluxes contribute to the inductive nature of the armature winding, resulting in the creation of a self-induced EMF known as reactance EMF. This reactance EMF leads the current Ia by 90°, meaning it's ahead of the current in terms of its phase. Considering these effects, the armature winding is often assumed to possess a leakage reactance, referred to as X_L. This concept helps explain how the armature winding behaves when dealing with these inductive effects caused by the leakage fluxes.

Voltage Drop due to Armature reaction, X_{ar}:
During no load condition, the magnetic flux in the air gap is generated solely by the rotor's Magneto Motive Force (MMF). However, under a load, the armature conductors carry a load current, leading to the creation of an armature flux. As a result, the air gap's magnetic flux changes compared to the no-load condition due to the influence of the armature flux.
Thus in the air gap, there are two distinct fluxes: one is the main flux or field flux, and the other one is the armature flux. This armature flux interacts with the magnetic flux produced by the rotor's poles. This interaction causes the resultant flux in the air gap to either decrease or increase in comparison to the main field flux in the alternator. This phenomenon, where the armature flux influences the main flux generated by the rotor's or field's magneto motive force, is termed armature reaction.

The armature reaction leads to a voltage drop in the armature winding, which needs to be taken into account. To deal with this, the armature winding is often given an artificial reactance known as "fictitious reactance" (X_{ar}). This value of X_{ar} is chosen so that when multiplied by the armature current (I_a), it accurately represents the voltage drop caused by armature reaction. This way, the impact of armature reaction on the machine's performance can be better managed and analyzed. Also, it's a significant consideration in alternator design and operation, as it affects the machine's performance and characteristics when it's under load.

It's important to recognize that both armature and main fluxes rotate at the same synchronous speed and direction. The impact of armature reaction is influenced by two main factors: the power factor of the load and the magnitude of the armature current. The power factor of the load determines whether the armature flux distorts, opposes, or assists the flux generated by the rotor's MMF. This phenomenon is essential to consider because it influences the machine's overall performance.

6.2 EFFECT OF LOAD POWER FACTOR ON ARMATURE REACTION

6.2.1 Unity Power Factor Load

When the load has a unity power factor (UPF), which means it's purely resistive, armature reaction affects the main field flux in a way that it doesn't weaken it but distorts it.

In the phasor diagram shown in Figure 6.2, there's a 90° phase difference between the field flux and the armature winding flux. This distortion due to armature reaction causes a slight drop in the terminal voltage in the alternator. For a unity power factor load, armature reaction has a cross-magnetizing effect.

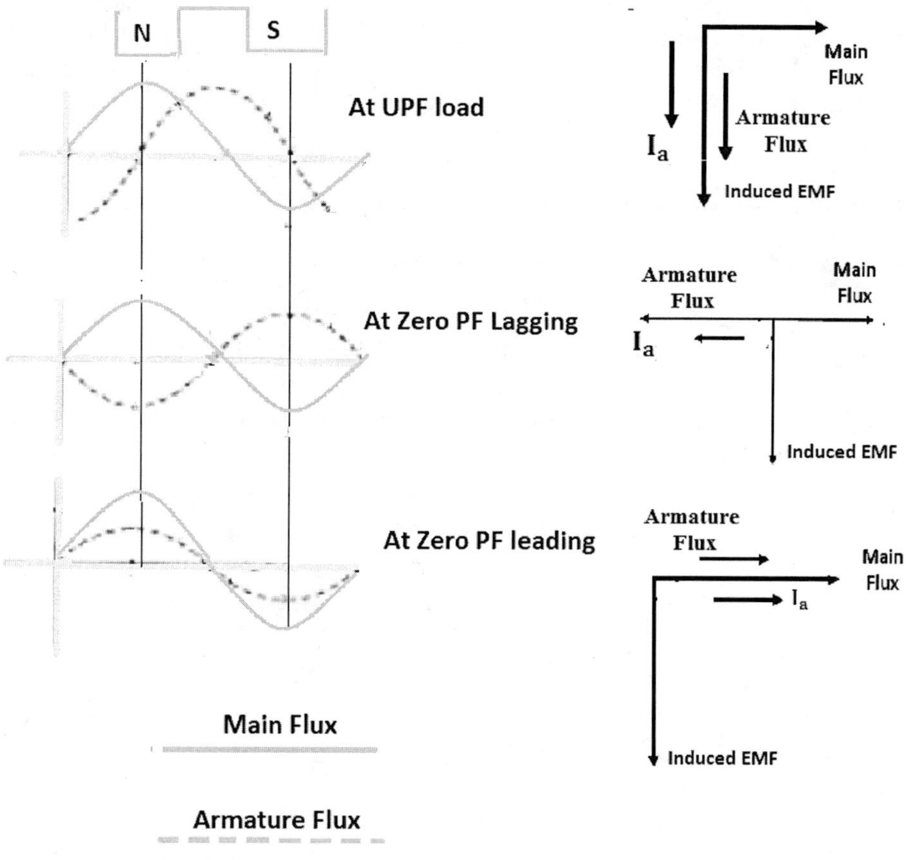

Fig.6.2: Effect of Load Power Factor on Armature Reaction

6.2.2 Zero Lagging Power Factor Load

When the load power factor is zero lagging, indicating a purely inductive load, armature reaction weakens the main field flux instead of distorting it. This weakening of the main field flux leads to a reduction in the generated electromotive force (EMF).In the vector diagram of Figure 6.2, you can observe that the armature and field winding fluxes are entirely opposite in direction. This effect of armature reaction, in this case, reduces the main flux and consequently decreases the EMF generation. For zero lagging power factor pure inductive loads, armature reaction acts in a demagnetizing manner. When the alternator supplies a lagging power factor load, the armature reaction is partly demagnetizing and partly cross magnetizing.

6.2.3 Zero Leading Power Factor Load

In situations where the load power factor is zero leading, meaning it's purely capacitive, armature reaction strengthens the main field flux without distorting it. This strengthening results in an increase in the generated EMF, boosting the alternator's performance.

In the vector diagram of Figure 6.2, you can observe that the armature and field winding fluxes are in the same direction, with 0° phase difference between the field flux and the armature winding flux. For zero leading power factor pure capacitive loads, armature reaction acts in a magnetizing manner. When the alternator supplies leading power factor load, then the armature reaction is partly cross magnetizing and partly magnetizing.

6.3 TERMINAL VOLTAGE EQUATION- SYNCHRONOUS IMPEDANCE

For alternator, the terminal voltage Vt can be written as

$$V_t = E_g - I_a R_a - jI_a X_L - jI_a X_{ar} \quad \text{-------(1)}$$

$$V_t = E_g - I_a R_a - jI_a (X_L + X_{ar})$$

The leakage reactance X_L and the armature reaction reactance X_{ar} may be combined to give synchronous reactance X_S.

$$V_t = E_g - I_a R_a - jI_a (X_s) \quad \text{-------(2)}$$

$$V_t = E_g - I_a (R_a + jX_s) \quad \text{-------(3)}$$

The total voltage drop in an alternator under load is= $I_a R_a + jI_a X_S = I_a(R_a + jX_S) = I_a Z_S$.

Z_s is known as synchronous impedance of the armature.

Thus, from equations (1) to (3), the terminal voltage becomes,

$$V_t = E_g - I_a Z_s \quad \text{-------(4)}$$

Where,

Eg - induced EMF,

Ia - armature current,

Ra - armature resistance,

XL - leakage reactance,

Xar - armature reaction reactance,

Xs - synchronous reactance

Zs - synchronous impedance

6.3.1 Phasor diagram of Alternator under various load conditions

The phase relationship between armature induced EMF, E_g due to field flux and the current flowing through the armature I_a depends upon the power factor of the load. The induced EMF E_g can be obtained from the terminal voltage V_t analytically using the equation,

$$E_g = V_t + I_a Z_s$$

$$E_g = V_t + I_a R_a + jI_a X_s \qquad \text{------- (5)}$$

Resistive Load:

The vector diagram given in Figure 6.3 is drawn to obtain the induced EMF, E_g from the terminal voltage V_t, considering the terminal voltage V_t as a reference phasor. When the load connected to the alternator is resistive, the armature current maintains the same phase as the terminal voltage (OP).The armature resistance drop I_aR_a (PQ) is due to the armature current, I_a and hence it always lies in phase with current I_a (i.e., PQ in phase with OP, and $||^{el}$ to PQ). Drop due to synchronous reactance jI_aX_S (QR) can be drawn from PQ taking 90° shift to get I_aZ_S. E_g is then obtained using Equation 5.

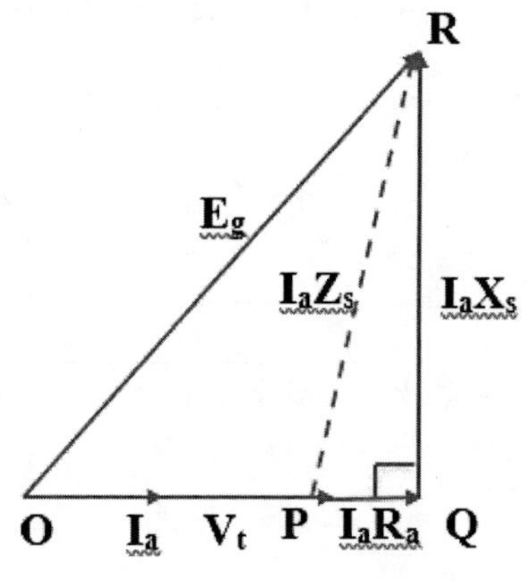

Fig.6.3: Induced EMF in the alternator for Resistive load - Phasor diagram

From the triangle OQR, the expression for induced EMF, E_g is given as,

$$OR^2 = OQ^2 + QR^2$$

$$E_g{}^2 = (OP + PQ)^2 + QR^2$$

$$E_g = \sqrt{(V_t + I_a R_a)^2 + (I_a X_s)^2}$$

Inductive Load: RL load

At a zero lagging power factor (purely inductive), the current I_a lags behind the voltage V_t(OP) precisely by 90°. However, for loads with lagging power factors, the current I_a lags behind the terminal voltage V_t by an angle $\emptyset$. The provided Figure 6.4 illustrates a phasor diagram for a lagging power factor scenario.

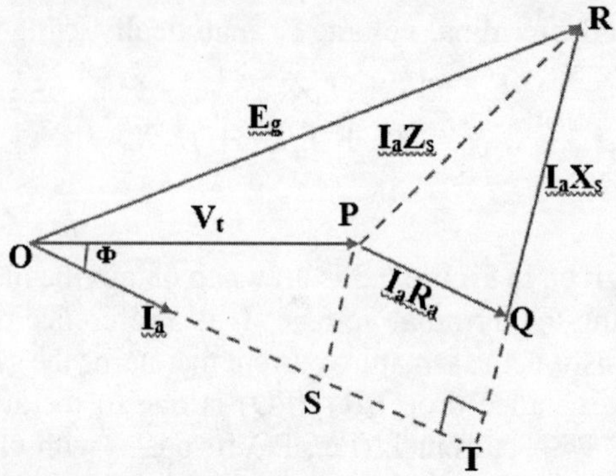

Fig.6.4: Induced EMF in the alternator for RL load - Phasor diagram

The armature resistance drop I_aR_a (PQ) is due to the armature current I_a and hence it always lies in phase with current I_a (i.e., ST in phase with OS, and $||^{el}$ to PQ). Drop due to synchronous reactance jI_aX_S is shown by QR, taking 90° shift from PQ to get I_aZ_S. E_g is then obtained using equation 5.Therefore, from the triangle ORT,

$$OR^2 = OT^2 + TR^2$$

$$E_g{}^2 = (OS + ST)^2 + (TQ + QR)^2$$

$$E_g = \sqrt{(V_t \cos\emptyset + I_aR_a)^2 + (V_t \sin\emptyset + I_aX_s)^2}$$

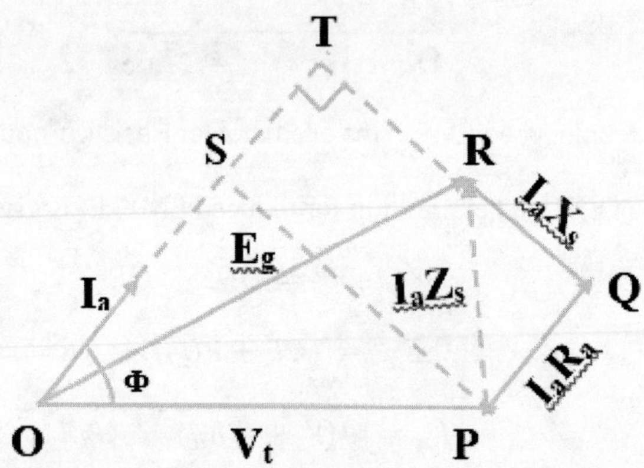

Fig.6.5: Induced EMF in the alternator for RC load - Phasor diagram

Capacitive Load: RC load

At a zero leading power factor (purely capacitive), the current I_a leads V_t exactly by 90°. In the case of a leading power factor load, the armature current I_a leads the terminal voltage V_t by an Angle □ as shown in Figure 6.5.

$$OR^2 = OT^2 + TR^2$$

$$E_g{}^2 = (OS + ST)^2 + (QT - QR)^2$$

$$E_g = \sqrt{(V_t \cos \emptyset + I_a R_a)^2 + (V_t \sin \emptyset - I_a X_s)^2}$$

Based on the information provided above, we can draw the following conclusions:

For cases of unity and lagging power factors, the product of I_a (armature current) and X_s (synchronous reactance) will have a positive sign. This is because at unity and lagging power factors, the armature reaction's effect on X_s is both demagnetizing and cross-magnetizing.

For a leading power factor, the product of I_a and X_s will have a negative sign. This means that the terminal voltage (V_t) is greater than the generated electromotive force (Eg) due to the magnetizing effect of armature reaction.

6.4 VOLTAGE REGULATION

Loading an alternator causes its terminal voltage to drop or rise depending on the (i) Magnitude of load and (ii) Nature of the load. The change in voltage from no load to full load is given by parameter called voltage regulation. The performance of the machine is better, if the voltage regulation value is small. Voltage regulation refers to the percentage increase in voltage that occurs when full load is disconnected, excitation being set to provide the normal voltage initially. It is expressed in percentage.

$$\text{Regulation} = \frac{E_{go} - V_t}{V_t} \text{x } 100\ \%$$

where,
V_t is the terminal voltage per phase;
E_{go} is the induced voltage per phase.

There are several methods in alternatorin order to estimate E_{go}. They are direct loading method, synchronous impedance method (EMF method), ampere turns method (MMF method), and Potiertri angle method(ZPF method). Direct loading method can be used, in case of small generators, whereas for large alternators, E_{go} cannot be determined by the direct loading method. Itis usually predetermined by any of the other different methods, applicability limited to non salient pole alternators.

6.4.1 Synchronous Impedance Method (EMF method)

This EMF method requires the following.

- Armature resistance, Ra,
- Open Circuit Characteristic (OCC), and
- Short Circuit Characteristic (SCC)

No load tests are conducted to find the amature resistance per phase, OCC upto 125% of rated voltage, and SCC for rated current.

- Armature Resistance (Ra):

The per-phase value of Ra can be directly determined using methods like the Wheatstone bridge, voltmeter, and ammeter approach. As a general rule, a value approximately 1.6 times the DC resistance is commonly used for calculations.

- Open Circuit Characteristic (OCC):

The OCC can be acquired by operating the machine under no-load conditions and noting the induced voltage values at the stator winding terminals as the field excitation current varies. This process continues until the voltage reaches 125% of the rated value.

- Short Circuit Characteristic (SCC):

The SCC is obtained by short-circuiting the armature windings through a low-resistance ammeter. The machine is then run at its rated speed, gradually increasing the field current until the armature current reaches the rated value.

6.4.2 Procedure to obtain OCC and SCC

For the given alternator motor coupled set, the ratings and the required apparatus details are provided in the Tables 6.1 and 6.2 respectively to conduct OC and SC tests on alternator.

Table 6.1: Rating of alternator and motor

ALTERNATOR		MOTOR
Power Rating:	5 kVA	7.5 kW
Rated Voltage:	415 V	415 V
Rated Current:	7.5 A	4.5 A
Rated Speed:	1500 rpm	1440 rpm
Excitation:	1.2 A	

The connections are to be given as shown in the circuit diagram in Figure 6.6. Keeping the motor field rheostat in the indicated position and with the TPSTS open, the motor supply is switched ON, by closing $DPSTS_1$.Motor is started using the 3-point starter by moving the handle from OFF to ON position and the motor is brought to its rated speed by adjusting the rheostat in the motor field circuit. Supply is switched on to the field winding of alternator by closing the $DPSTS_2$. The procedure to plot OCC and SCC is given in below.

Table 6.2: Apparatus required to conduct OC and SC Tests on Alternator

S. No	Name of the Apparatus	Range	Type	Quantity
1	Rheostat	200ohms/3A		2 Nos
2	Voltmeter	(0-600V)	MI	1 No
3	Ammeter	(0-10A)	MI	1 No
4	Ammeter	(0-2A)	MC	1 No
5	Tachometer	-		1No
6	DPSTS and TPSTS	-		1 No each
7	Connecting wires & Fuse	-		As Required

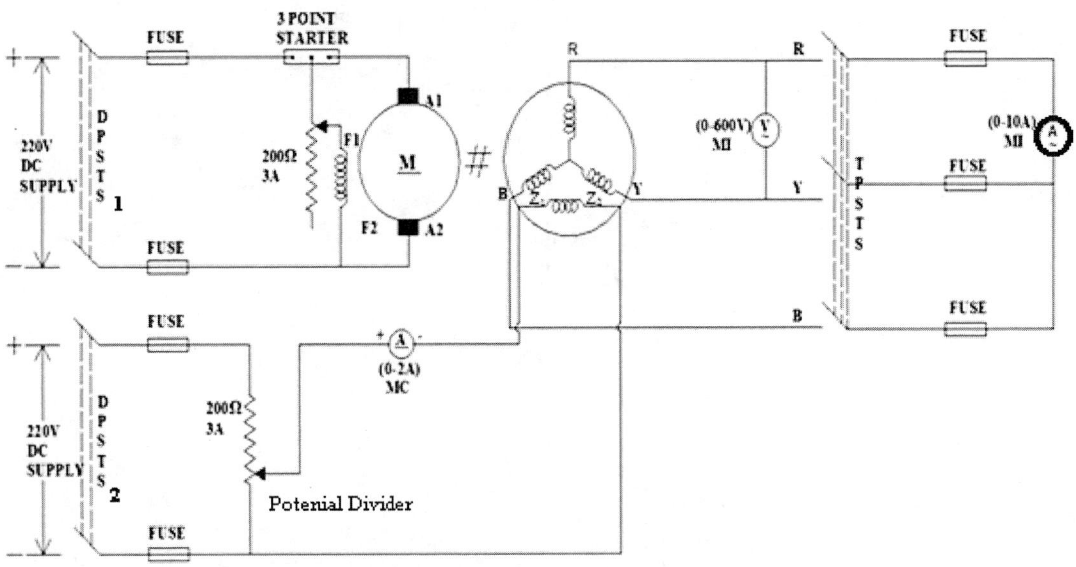

Fig.6.6: Circuit to conduct OC and SC tests on alternator

OC Test Procedure

1. Using the 200 ohm potential divider, current in field circuit is increased in steps of 0.1A and at each step the alternator induced voltage indicated by voltmeter and the corresponding field current (I_f) are noted.
2. This procedure is continued until the alternator voltage is 125% of its rated voltage.
3. Using the data, OCC curve can be plotted taking the field current long x-axis, and induced EMF along y-axis, E_{go} Vs I_f.

S. No.	I_f (A)	$E_{go(1-1)}$ (V)	$E_{go\ (ph)}$(V)

S.C. Test Procedure

1. After Completing O.C. Test, the potential divider is brought to its minimum position.
2. The alternator terminals are then short circuited by closing TPST switch through an ammeter.
3. The rated current is made to flow through the armature of the stator windings by carefully adjusting 220 ohms potential divider from the minimum position.
4. Field current is noted from the ammeter connected in the field winding, for the rated full load current in the armature winding ammeter.
5. Using the data, graph can be plotted against I_{sc} Vs I_f.
6. Figure 6.7 gives the OCC and SCC.

S. No.	I_f (A)	I_{SC} (A)

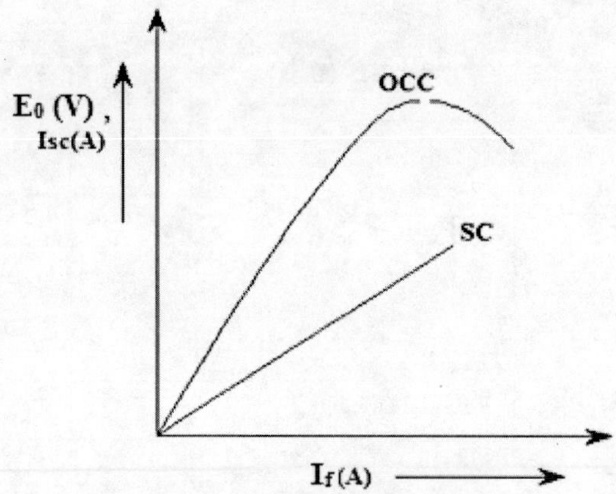

Fig.6.7: OCC and SCC of alternator

After completing these OC and SC tests experimentally, synchronous impedance, and synchronous reactance are calculated using the formulae. The steps involved in the voltage regulation calculation of EMF method are as follows.

Steps involved in Synchronous Impedance calculation and Regulation predetermination:

1. OCC curve is drawn with induced EMF for varying field current from the OC test data. E_{go} Versus I_f.

2. To create the short circuit characteristic SCC (straight line), origin is connected with a point corresponds to the field current required to circulate the full load rated current.

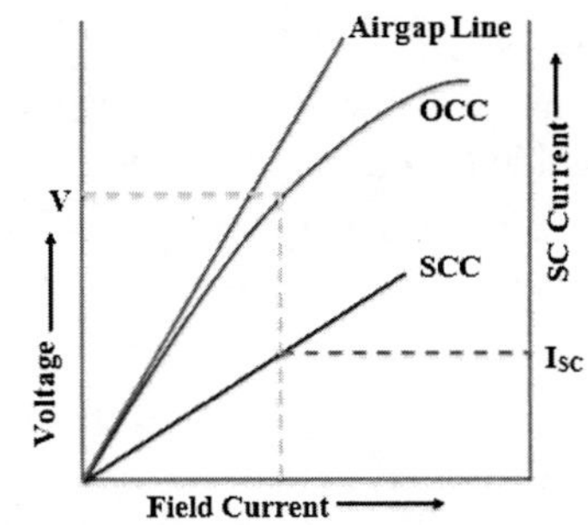

Fig.6.8: Extraction of Open circuit voltage and short circuit current

3. To Calculate the Synchronous Impedance (Zs):

Under identical field current conditions (from Figure 6.8 - graph),

$$Z_s = \frac{(Open\ circuit\ voltage\ per\ phase)}{(Short\ circuit\ current\ per\ phase)}$$

$$Z_s = \frac{V_{OC}}{I_{SC}}$$

$$\text{Synchronous Reactance}, X_S = \sqrt{{Z_S}^2 - {R_a}^2}$$

$$E_{go} = \sqrt{(V_t \cos\emptyset + I_a R_a)^2 + (V_t \sin\emptyset \pm I_a X_s)^2}$$

Where,

E_{go} = no load Induced EMF per phase,
V_t = Rated voltage per phase,
R_a = Armature resistance in ohms,
I_a = Armature current in Amps,
'+' for lagging pf. load.
'-' for leading pf. load.

4. Regulation is calculated as

$$\%\ \text{reg} = \frac{E_{go} - V_t}{V_t}$$

6.5 MAINTENANCE OF ALTERNATORS

An alternator is an electro-mechanical device with stator, rotor winding and an external DC field excitation. Alternator generates power on coupling with a prime mover. Maintenance of alternators is essential to ensure their proper functioning and longevity. Regular maintenance helps prevent breakdowns, improves efficiency, and reduces the risk of costly repairs.

Before starting any maintenance work on the alternator, all safety precaution should be taken. Also, the manufacturer's guidelines and recommendations for maintenance are to be followed. Regular maintenance not only prolongs the life of the alternator but also ensures the overall health and reliability of the electrical system in your vehicle or machinery.

Regular inspections:

Visual inspections of the alternator is necessary to check for any signs of damage, wear, or loose connections. Look for frayed wires corrosion, and broken parts. Any irregularities, strange noises, or other signs of problems have to be addressed before they escalate.

Some of the inspection key points are as follows.

- Clean the alternator ventilation passage.
- Check the Insulation resistance of stator and rotor winding.
- Ensure that the alternator is securely mounted in place. Loose mounts can cause vibrations, leading to damage over time.
- Air gap between stator and rotor to be checked and maintained between 1.5 to 2 mm.
- Slip rings to be checked for even wear down to be renewed if required.
- Automatic Voltage Regulator to be checked and cleaned off oil and dust.
- A vacuum cleaner can be used to remove dust accumulated in the inner parts of alternator.
- All the connection in the terminal box to be tightened properly.
- Forced Ventilation around alternator must be maintained all the time.
- The foundation bolts of the alternator to be checked for tightness.
- Check the tension of the alternator belt regularly. It should have the correct amount of tension to drive the alternator properly. Too loose or too tight can cause issues.

After regular maintenance is performed, the following are the major tests to be conducted periodically.

Cooling system:

Alternator's cooling system should be working efficiently. Overheating can lead to damage and premature failure. This testing will help to address the issues in thermal aspects.

No-load Test:

A no load test should be carried out and general condition such as noise, temperature, voltage generated etc. of the alternator should be observed and noted.

Regular load testing:

Load tests on the alternator determine its ability of the alternator to handle electrical loads. This will help to locate any potential load issues.

UNIT SUMMARY

The aim of this chapter is to analyse the factors affecting the terminal voltage of the alternator, induced EMF of the alternator on load, effect of load power factor on armature reaction, phasor diagrams of induced EMF's of the alternator for various load power factor and voltage regulation of the alternator. Also, for the synchronous impedance method, this chapter explains how to predetermine the voltage regulation of the alternator experimentally with open circuit and short circuit tests.

EXERCISES

1. Explain the power developed by a synchronous generator.
2. Explain the parallel operation of alternator.
3. Explain the voltage regulation of the alternator by MMF method.
4. Explain the voltage regulation of the alternator by Potier triangle method.
5. The OC and SC test results of the alternator is given as follows.

Open circuit voltage - $E_{go\,(\text{line to line})}$
Field current - I_f

S. No.	I_f (A)	$E_{go(l-l)}$ (V)
1	16	3100
2	25	4900
3	37.5	6600
4	50	7500
5	70	8300

Ignore the armature resistance and leakage reactance. Determine the regulation at 0.8pf using synchronous impedance method. A field current of 20A is found required for the circulation of full load current on short circuit test.

Multiple Choice Questions

1. The magnitude of leakage flux in alternator depends on
   a. Air gap thickness
   b. The magnitude of armature current
   c. Phase angle between terminal voltage and armature current
   d. Both a and b

2. For zero factor leading, the effect of armature reaction in an alternator on the main flux is
   a. Magnetizing
   b. Demagnetizing
   c. Cross-magnetizing
   d. None of these

3. Which of the following methods would give higher than actual value of regulation of an alternator?
   a. MMF method
   b. EMF method
   c. ZPF method
   d. ASA method

4. A polyphase field is
   a. Pulsating and rotating
   b. Pulsating and stationary
   c. Constant in amplitude and rotating at synchronous speed
   d. Constant in amplitude and stationery in space

5. The necessary condition for parallel operation of two alternators is
   a. Terminal voltage should be the same
   b. Phase sequence should be the same
   c. Frequency should be the same
   d. All the these

Answers of Multiple Choice Questions

1. a. Magnetizing
2. a. Magnetizing
3. b. EMF method
4. c. Constant in amplitude and rotating at synchronous speed
5. d. All of these

Short and Long Answer Type Questions

Long Answer Questions with Answers

1. **Draw the equivalent circuit of the alternator that is in delta connection.**
   For delta connection the equivalent circuit of the alternator is given below.

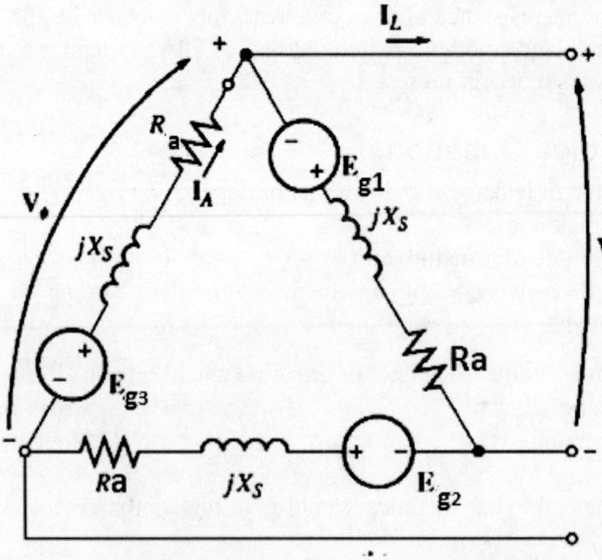

Fig. 6.9: Equivalent circuit of the alternator in Delta connection

2. Draw the equivalent circuit of the alternator that is in star connection.

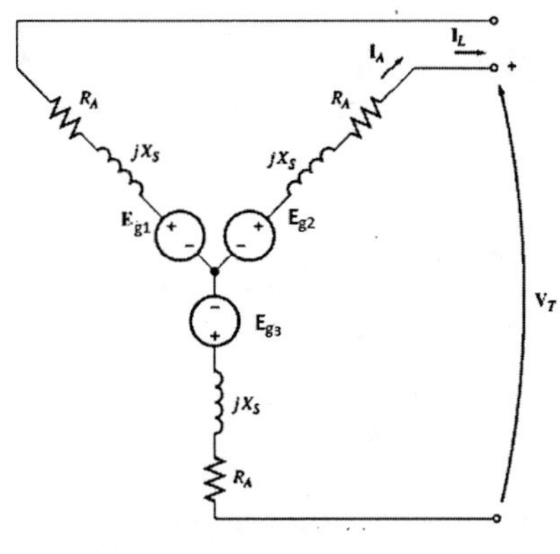

Fig. 6.10 Equivalent circuit of alternator in star connection

When an alternator is put under load, the entire induced electromotive force (EMF), ie., **Eg** doesn't fully appear across the terminals of the output, $\mathbf{V_t}$. As the load on an alternator change, the terminal voltage of the alternator also changes. This variation in terminal voltage is caused by different factors, including voltage drops due to the armature's resistance (Ra), armature leakage reactance (X_L), and the influence of the armature reaction.

For alternator, the terminal voltage Vt can be written as

$$V_t = E_g - I_aR_a - jI_aX_L - jI_aX_{ar} \qquad \text{-------} \quad (1)$$

$$V_t = E_g - I_aR_a - jI_a(X_L + X_{ar})$$

The leakage reactance X_L and the armature reaction reactance X_{ar} may be combined to give synchronous reactance X_S.

$$V_t = E_g - I_aR_a - jI_a(X_s) \qquad \text{-------} \quad (2)$$

$$V_t = E_g - I_a(R_a + jX_s) \qquad \text{-------} \quad (3)$$

The total voltage drop in an alternator under load is $= I_aR_a + jI_aX_S = I_a(R_a + jX_S) = I_aZ_S$.

Z_S is known as synchronous impedance of the armature.

Thus, from equations (1) to (3), the terminal voltage becomes,

$$\boldsymbol{V_t = E_g - I_aZ_s} \qquad \text{-------} \quad (4)$$

where,

Eg is the induced EMF,

Ia is the armature current,

Ra is the armature resistance,

XL is the leakage reactance,

Xar is the armature reaction reactance,

X_S is the synchronous reactance

Z_S is the synchronous impedance

Hence the equivalent circuit of the alternator in Figure 6.1 becomes that in the Figure 6.7. For delta connection the equivalent circuit of the alternator is given below.

Short Questions:

1. Define synchronous reactance.
2. Write down the equation for the terminal voltage of the alternator on load.
3. Does the synchronous machine has slip speed?
4. What is the effect of lagging load power factor on armature reaction?
5. Define voltage regulation of the alternator.
6. What are the methods to determine voltage regulation?
7. Write down the formula to determine the synchronous impedance.
8. What are the factors that cause variations in the terminal voltage of the loaded alternator?

Long Questions:

1. Explain the vector diagrams for the loaded alternator.
2. Explain about alternator maintenance scheduling considering the economic power generation.
3. Explain the factors to be considered for the application of alternator in the hydroelectric power generating plants.
4. Explain the site location and limitations for deploying the alternator in the thermal electric power generating plants.
5. Derive the expression for terminal voltage for the loaded alternator under various power factor load.
6. Explain the effect of armature reaction on the alternator.
7. Derive the reactive power output per phase of the alternator.
8. Detail the effect of various load power factor on armature reaction?

Numerical Problems

1. An alternator has a rated voltage of 11kV and is operating at 0.8 power factor lagging. If the load current is 100A, calculate the terminal voltage. Consider the synchronous reactance as 2 Ω.
2. A 3-phase alternator is loaded to 80%80% of its rated kVA. If the field excitation is reduced by 10%, calculate the new terminal voltage. Assume a power factor of 0.85 lagging.
3. An alternator has an armature resistance of 0.5Ω and a leakage reactance of 2Ω. If the alternator is delivering 1000A to a load with a power factor of 0.9 lagging, calculate the armature voltage drop.
4. An alternator operates at a power factor of 0.9 leading. If the synchronous reactance is 3Ω and the load current is 800A, calculate the armature reaction voltage.

5. An alternator has a rated voltage of 20kV and a synchronous reactance of 5Ω. If the full-load current is 1000A with a power factor of 0.9 lagging, calculate the voltage regulation using the direct loading method.
6. During routine maintenance, the insulation resistance of an alternator is measured and found to be 2MΩ. If the acceptable minimum insulation resistance is 1MΩ, evaluate the condition of the insulation.

PRACTICAL

1. Set up a small alternator and load it progressively. Measure the terminal voltage at different loads and analyze the impact of factors such as load current, power factor, and field excitation on terminal voltage.
2. Introduce variations in field excitation and observe the effect on terminal voltage. Discuss the importance of maintaining optimal excitation for stable voltage output.
3. Connect a load bank to the alternator and vary the load resistance. Measure the voltage drop across the armature resistance and leakage reactance to illustrate how these factors influence the overall performance.
4. Use circuit simulations or a physical model to showcase the concept of armature resistance and leakage reactance drops. Discuss how these drops affect the efficiency and voltage regulation of the alternator.
5. Vary the power factor of the load and observe the changes in armature reaction. Use instruments to measure the armature reaction at different power factors and discuss its impact on the alternator's performance.
6. Perform routine maintenance tasks on an alternator, such as cleaning, inspecting brushes and slip rings, checking insulation resistance, and lubricating moving parts. Discuss the importance of regular maintenance in ensuring the reliable operation of alternators.

KNOW MORE

1. Armature reaction of the alternator
   https://youtu.be/8HFa_YjZIFc?si=pL5ygSrgYHO5VqKO
2. Terminal voltage under load and Voltage regulation of alternator
   https://youtu.be/a5usKz1YRsY?si=osNmUQwtmm0Ke7-o

REFERENCES AND SUGGESTED READINGS

1. P.S. Bimbhra, Electric Machines, Khanna Book Publishing Co., New Delhi (ISBN: 978- 93-6173- 294)

2. Mittle, V.N. and Mittle, Arvind., Basic Electrical Engineering, McGraw Hill Education New Delhi, ISBN :9780070593572

3. Kothari, D. P. and Nagrath, I. J., Electrical Machines, McGraw Hill Education. New Delhi, ISBN:9780070699670

4.Bhattacharya, S. K., Electrical Machines, McGraw Hill Education, New Delhi, ISBN:9789332902855

5. Theraja, B.L., Electrical Technology Vol-II (AC and DC machines), S.Chand and Co. Ltd., New Delhi, ISBN : 9788121924375
6. Hughes E., Electrical Technology, ELBS
7. Cotton H., Electrical Technology, ELBS

Dynamic QR Code for Further Reading

7 SYNCHRONOUS MOTOR

UNIT SPECIFICS

Through this module we have discussed the following aspects:

- *Familiarizing the working of Synchronous motor;*
- *Familiarizing the phasor diagram of Synchronous motor;*
- *Different torques in a Synchronous motor;*
- *Power flow equations of Synchronous motor;*
- *Synchronous motor on load with constant excitation;*
- *Effect of excitation at constant load;*
- *V-Curves and Inverted V-Curves;*
- *Hunting and Phase swinging;*
- *Methods of Starting of Synchronous Motor;*
- *Losses in synchronous motors and efficiency;*
- *Applications of Synchronous motor.*

The photos of practical motor are shown to detail the various parts of the machine. The real time applications of the topics are discussed for inducing further curiosity and creativity.

Besides giving a large number of multiple-choice questions as well as questions of short and long answer types marked in two categories following lower and higher order of Bloom's taxonomy, assignments through a number of numerical problems, a list of references and suggested readings are given in the unit so that one can go through them for practice. It is important to note that for getting more information on various topics of interest some QR codes have been provided in different sections which can be scanned for relevant supportive knowledge.

After the related practical, based on the content, there is a "Know More" section. This section has been carefully designed so that the supplementary information provided in this part becomes beneficial for the users of the book. This section mainly highlights the initial activity, examples of some interesting facts, analogy, history of the development of the subject focusing the salient observations and finding, timelines starting from the development of the concerned topics up to the recent time, applications of the subject matter for our day-to-day real life or/and industrial applications on variety of aspects, case study related to environmental, sustainability, social and ethical issues whichever applicable, and finally inquisitiveness and curiosity topics of the unit.

RATIONALE

Synchronous motors are used in a variety of industries, including oil and gas, industrial & manufacturing, metals & mining, and others. The major market, which includes the use of synchronous motors for pumps, extruders, and compressors, is captured by the industrial & manufacturing industry. This chapter introduces the students to the construction and working principle of the synchronous motor. Also briefly discussed on the methods of Starting of Synchronous Motor; losses in synchronous motors and efficiency and applications of Synchronous motor.

PRE-REQUISITES

Machines I: Sem III

Physics: Electromagnetism (Class X)

UNIT OUTCOMES

List of outcomes of this unit is as follows:

U7-O1: Understand construction of the synchronous motor.

U7-O2: Able to understand the constructional difference in cylindrical and projected pole.

U7-O3: Feel the synchronisation of the stator and rotor.

U7-O4: Why self-starting is not possible in synchronous motor.

U7-O5: Able to get to know about the zero slip.

Unit-7 Outcomes	EXPECTED MAPPING WITH COURSE OUTCOMES (1- Weak Correlation; 2- Medium correlation; 3- Strong Correlation)				
	CO-1	CO-2	CO-3	CO-4	CO-5
U7-O1	-	-	-	3	-
U7-O2	-	-	-	3	-
U7-O3	-	-	-	3	-
U7-O7	-	-	-	3	-
U7-O5	-	-	-	3	-

7.1 INTRODUCTION

The Electric motors which are advanced one as well as the Modern electrical power transmission are based on 3-Φ electric power system, which was invented in the year 1885. There are several renowned individuals who all had significant involvement in the inventions of three phase power system (Dobrowolsky, Haselwander, Tesla and Wenström, Dolivo, Ferraris, Bradley,). Application like Electric cars and robots which are highly dynamic mostly uses three phase synchronous motor. In 1887, Friedrich August Haselwander invented the first Synchronous Machine. The first synchronous generator which is of three phase type with salient poles was built by Friedrich August Haselwander. However, the German Post the postal service known as German Post forbids the use of his device out of concern for the disruption of telegraph wires. Haselwander's patent applications failed as well. Haselwander's synchronous motor made in1887 is shown in Fig.7.2. The practical synchronous motor used in industry is shown in Fig.

Fig.7.1: Haselwander's synchronous motor made in1887

The advancements in the technology have resulted in inventions and innovations such that the Haselwander's synchronous motor practically used in practice. The rotor of a synchronous motor which is considered to be an AC motor revolves at the same rate as the rotating magnetic field (RMF). The speed at which magnetic field of stator revolves is known as synchronous speed which is a function of supply frequency. So this Machine is known as synchronous motor. In synchronous motor the rotor and the supply current frequency are synchronised. The rotor of a synchronous motor has either permanent magnets or field windings that are energised by an external source rather than relying on the induced rotor current. RMF is produced by the stator in synchronous motors. Synchronous speed is the rate at which the rotor's magnetic field rotates while magnetically locking with the rotating RMF.

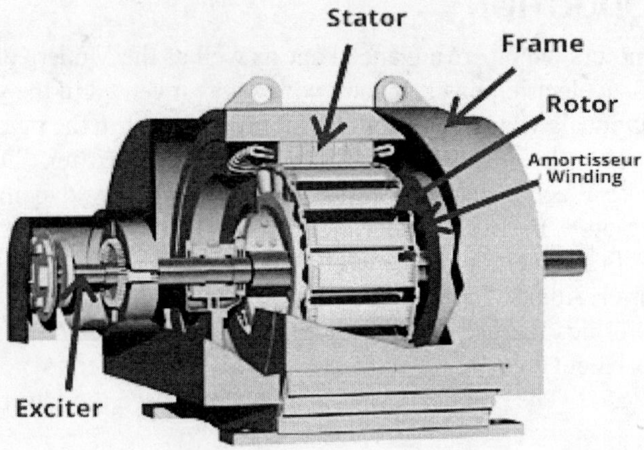

Fig.7.2: Synchronous motor

Because of its many advantages, such as power aspect correction for energy conservation and excellent efficiency, synchronous motors are more widely used in the industrial sector. Synchronous motors are frequently used in industry, where they may be used as generators or motors depending on the sector. To provide more consistency, it often operates at a synchronous tempo, which is a consistent speed with predictable frequency. Due to their outstanding performance, minimal power loss, low maintenance requirements, and adaptability to different plant layouts, synchronous motors are becoming more and more popular.

7.2 CONSTRUCTION OF SYNCHRONOUS MOTOR

Synchronous generator and Synchronous Motor have similar construction. Synchronous motor has two main parts namely Stator and Rotor. Stationary part of Synchronous Motor is known as stator. For reducing the eddy current loss and hysteresis loss core of the stator is made up off thin laminated sheets made from cast iron or steel having better magnetic property. The three-phase alternating stator field winding, also known as the armature winding, is held in axial slots in the core. When armature winding of stator is connected to three phase power supply, rotating magnetic field (RMF) is generated.

The synchronous motor's rotating part is called the rotor. Rotor has slots for holding field winding and is of cylindrical shape. It is responsible for generating the magnetic field or poles. DC source like a small DC generator is connected to the shaft of rotor can be used to energise the rotor through brush assembly and slip rings. Hence DC sources are used for excitation.

Salient Pole Rotor

The term 'salient' means 'pointing outward'. The salient Pole rotor has protruding or projecting poles toward the armature winding. The rotor core is made of a laminated steel sheet for reduction in hysteresis and Eddy current. The field windings are wounded around each pole.

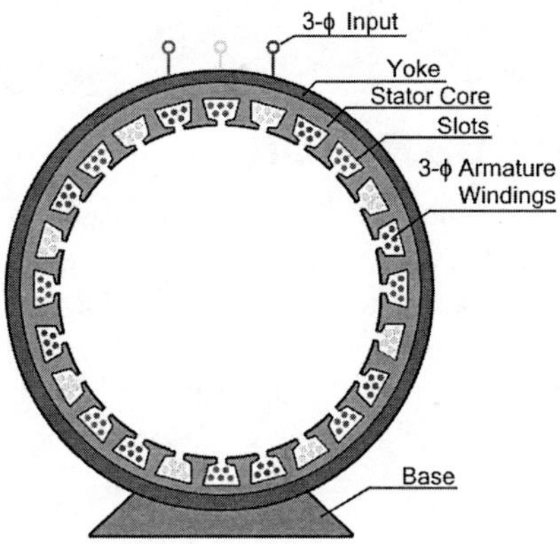

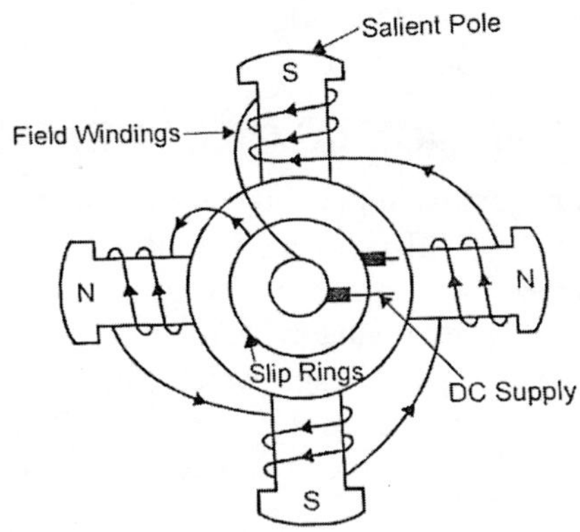

Fig.7.3: Stator of the Synchronous motor **Fig.7.4:** Salient pole rotor of Synchronous motor

Generally, in salient Pole type SM the number of poles is large; hence it is not suitable for applications with high-speed as high windage losses occur. This type of rotor is used in low and medium-speed synchronous motors. As the number of poles are more the diameter is large and axial length is small. Slip rings and brush assemblies are utilised to create an electrical connection between a machine's rotating part and stationary part. To energise the field winding through DC excitation salient pole type can be used.

Non-Salient Pole or Cylindrical Rotor

Non-salient pole rotor is also called cylindrical rotor because of its cylindrical shape and is made of laminated steel. Field windings are slotted in the core of the rotor that is wedge-shaped to keep them from being pulled out.

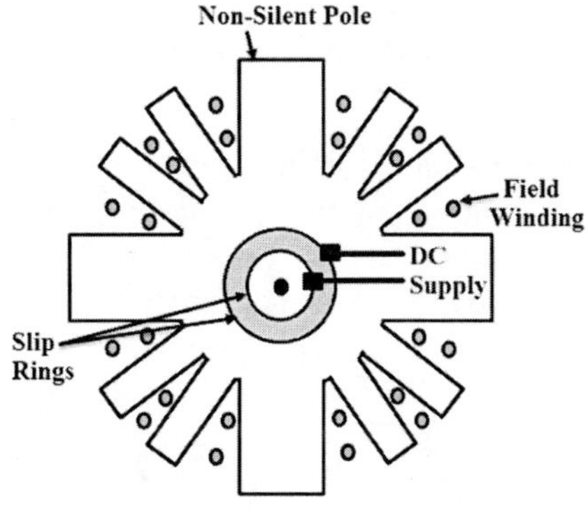

Fig.7.5: Cylindrical rotor of Synchronous motor

While the core's unslotted portion becomes magnetic pole In comparison to salient pole type rotor, costlier, less poles, a smaller diameter, and a longer axial length are all characteristics of the rotor this type of rotor. This type of rotor design helps in robustness flux, mechanical strength etc. and distribution of flux is uniform.

7.3 WORKING PRINCIPLE OF SYNCHRONOUS MOTOR

The Synchronous motors have the drawback of not being self-starting. An induction motor-like rotating magnetic field (RMF) with a constant magnitude and revolving at synchronous speed is created when a three-phase balanced voltage is given to the stator winding. This RMF can be predicated on two poles revolving synchronously in a clockwise orientation. Fig.7.7(a) depicts the Ns and Ss stator poles. The stator, which contains Ns at the top of the structure and Ss at the bottom, along with the rotor poles depicted in Fig.7.7(a), generate a repulsive force throughout the first half cycle. The rotor typically rotates anticlockwise.

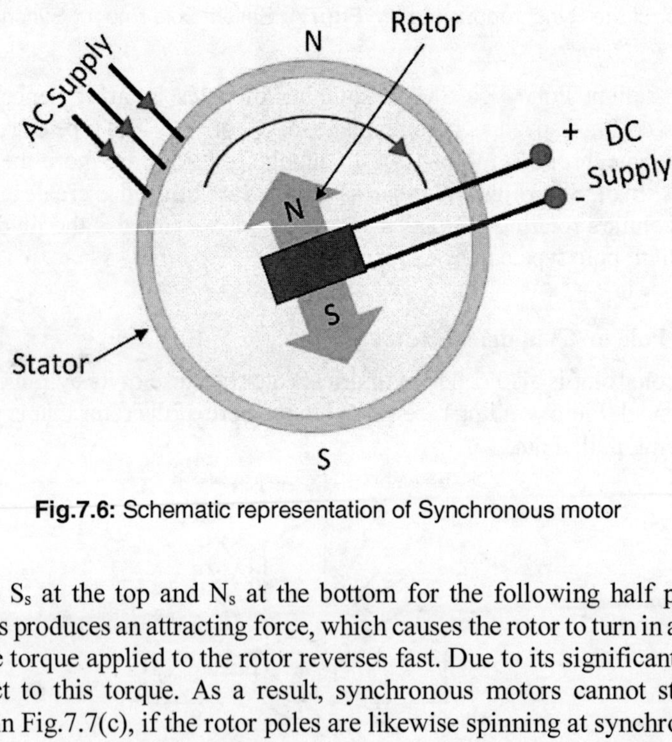

Fig.7.6: Schematic representation of Synchronous motor

The stator has S_s at the top and N_s at the bottom for the following half period, as shown in Fig.7.7(b). This produces an attracting force, which causes the rotor to turn in a clockwise manner. As a result, the torque applied to the rotor reverses fast. Due to its significant inertia, the rotor is unable to react to this torque. As a result, synchronous motors cannot start themselves. As demonstrated in Fig.7.7(c), if the rotor poles are likewise spinning at synchronous speed using a pony motor, the rotor poles will have switched positions by the time the stator poles do. The stator poles pull the rotor poles towards them and the two moves together. So, the magnetic locking concept underlies how the synchronous motor operates. A motor that operates synchronously is a doubly excited machine, meaning that in order to produce synchronism; it needs AC and DC power for both the stator and the rotor.

The windings of the stator receive a three-phase AC to produce RMF. The rotor and stator are intended to have an equal number of poles.

Synchronous speed is the rate of rotation of these poles that is in time with an input frequency f. It is provided by

$$N_s = 120f / P \qquad \text{-------(7.1)}$$

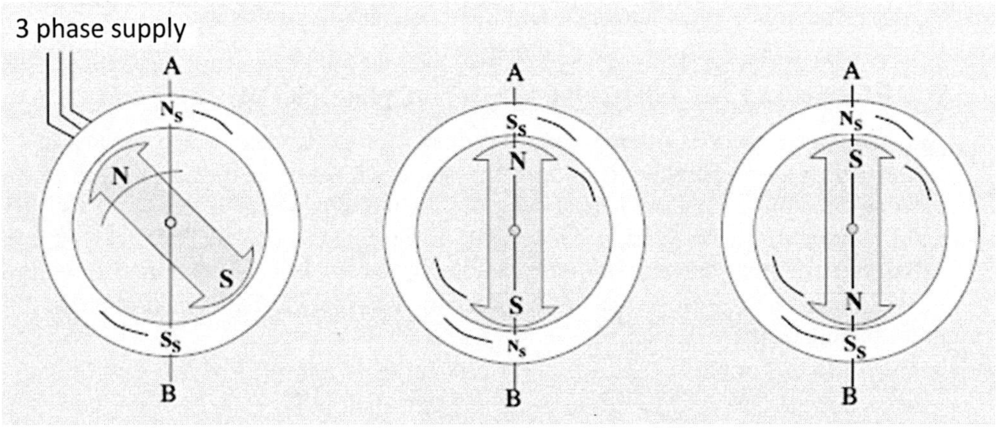

Fig.7.7(a) Fig.7.7(b) Fig.7.7(c)

Fig.7.7: Stator and rotor poles representation of Synchronous motor

The windings of the rotor receive a DC source in order to provide a constant magnetic field. The magnetic field of the rotor remains constant since the DC source delivers continuous current. At the opposing ends of the rotor, magnetic poles are produced. When the rotor reaches the synchronous speed, its poles begin to revolve at the same rate as the stator's RMF.

There will be no load torque if the shaft rotates substantially the same rate as the stator RMF. The poles of the stator and rotor line up. The rotor begins oscillating around the new equilibrium position when a mechanical force is applied; this behaviour is referred to as "hunting." The rotor begins to produce torque as it trails the stator RMF by a few degrees. The angle in between them grows as the load increases until the rotor field trails RMF by 90°. The motor now delivers its breakdown torque, which is the highest amount of torque. The motor stalls if the load is too great.

7.4 FEATURES OF SYNCHRONOUS MOTOR

The characteristics of synchronous motors are as follows.

- Synchronous motors cannot naturally start themselves. To synchronise with the supply frequency, the rotor must be accelerated to synchronous speed.

- Only the input supply's frequency affects its speed. The synchronous motor's speed is managed using a variable frequency drive (VFD).
- The load has no effect on its speed. As a result, any change in the load has no effect on the synchronous motor.
- The torque increases as the load does. Any torque increase that exceeds the breakdown torque will cause a synchronous motor to stall.
- The operation of a synchronous motor is either at synchronous speed or not at all.
- Leading and trailing power factors are both possible operating modes for synchronous motors. As a result, they are employed in industries to enhance power factors.

7.5 EQUIVALENT CIRCUIT OF SYNCHRONOUS MOTOR

A synchronous motor is often a doubly excited system as well, which implies that the motor rotates under two separate excitations. The stator winding receives a three-phase AC supply, creating a rotating magnetic field. Similar to that, the rotor winding receives a DC supply. Back e.m.f. is produced within the armature windings whenever the rotor rotates with synchronous speed. Keeping that in mind, the comparable circuit is sketched to illustrate the synchronous motor. The circuit representation of one phase for a cylindrical-type synchronous motor is shown in the picture below.

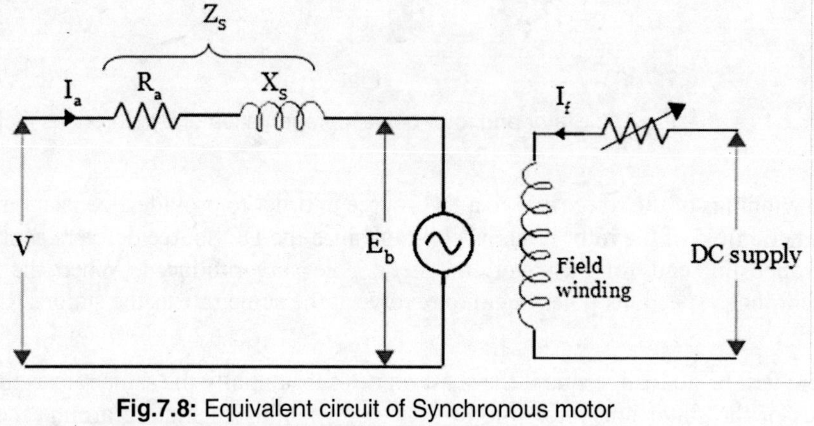

Fig.7.8: Equivalent circuit of Synchronous motor

In the equivalent circuit,

E = Excitation Voltage or E_f

I_f = Field current

V = It is the armature's applied terminal phase voltage.

I_a = The amount of armature current per phase the motor draws from the source

R_a = Effective armature resistance for each phase

X_S = Synchronous reactance per phase of the motor stator armature winding

Z_S = Impedance per phase of the armature

$\emptyset$ = Phase angle between V and I_a

$\cos \emptyset$ = Power factor

δ = torque angle= Phase difference between E_f and V

I_aR_a = It is the armature resistance, or voltage drop per phase.

I_aX_s = Reactive voltage drop per phase due to armature reactance and armature reaction effects

$$Z_s = R_a + jX_s \quad \text{-------(7.2)}$$

For a synchronous motor

$$V = E_f + I_aZ_s$$

$$V = E_f + I_a(R_a + jX_s)$$

$$E_f = V - I_aR_a - jI_aX_s \quad \text{-------(7.3)}$$

Phasor diagram of Cylindrical Rotor Synchronous Motor

A three-phase cylindrical rotor synchronous drive may run at lagging, unity, or leading power factors. In light of this, the phasor diagram is created using the aforementioned formulae. Understand how a synchronous motor reacts under no-load and load circumstances before reading the section.

(i) Phasor Diagram for Lagging Power Factor

Consider that the supplier is providing a trailing current to the synchronous motor. By taking equation (7.3) into consideration, a phasor diagram is created. In order to build that phasor diagram, the supply voltage's phasor value is used as a guide, with the result that OA = V. The armature current trails the input voltage; as a result, the phasor of the voltage V and the armature current phasor Ia follow each other by a phase angle φ.

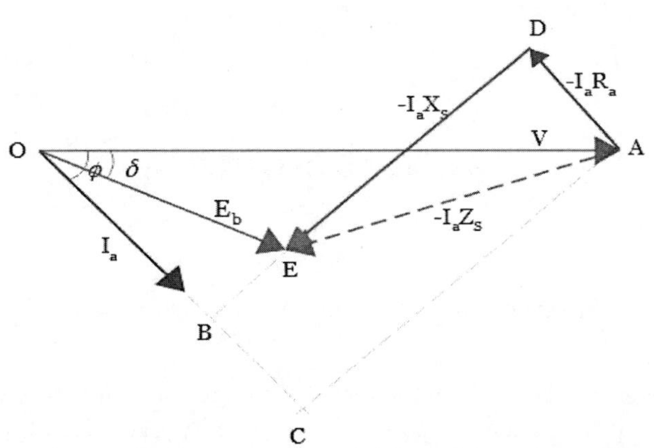

Fig.7.9 (a): Phasor diagram of Cylindrical Rotor Synchronous Motor for Lagging Power Factor

Draw the voltage drop phasor (AD = $-I_aR_a$), starting from point A, in the opposite direction from the current phasor. The voltage drop resulting from synchronous reactance draws perpendicular to the I_aR_a phasor from the extremity that defines this phasor (from point D) (DE = $-I_aX_s$). Join AE to obtain the voltage drop overall caused by total impedance ($-I_aZ_s$). Thus, $E_b = V - I_a R_a - jI_a X_s = V - I_a Z_s$. Join OE, which indicates the electromagnetic field that is generated per phase for the phasor E_b. Power angle or torque angle δ refers to the phase distinction between V and E_b.

We can also draw the phasor diagram considering the other equation, $V = E_b + I_a R_a + jI_a X_s$

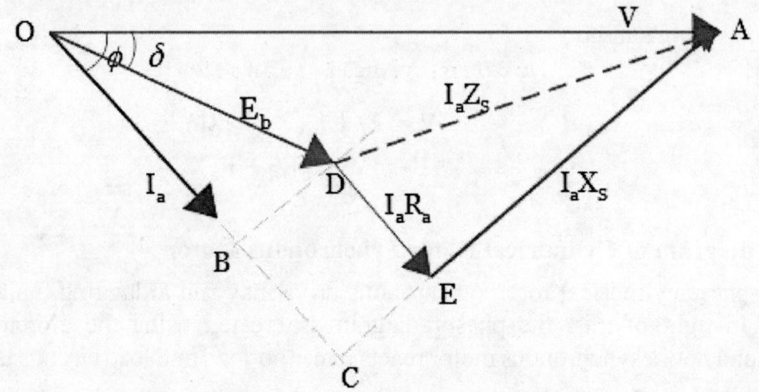

Fig.7.9 (b): Phasor diagram of Cylindrical Rotor Synchronous Motor for Lagging Power Factor

(ii) Phasor Diagram for Unity Power Factor

Consider that the synchronous machine is obtaining its current from the source with a power factor of unity. Equation (3) is taken into account while drawing the phasor diagram.

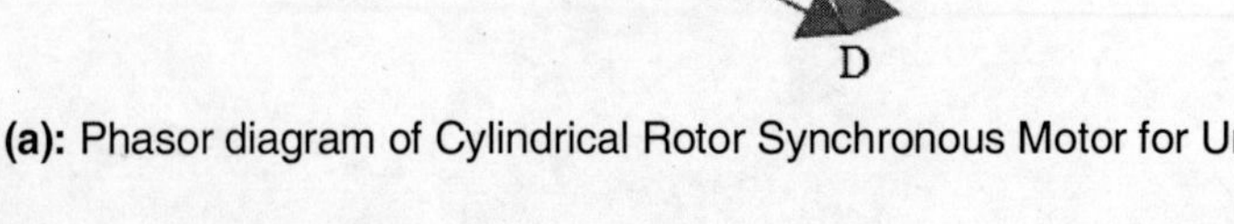

Fig.7.10 (a): Phasor diagram of Cylindrical Rotor Synchronous Motor for Unity Power Factor

Create a reference phasor using the terminal voltage so that OA = V. The armature current and voltage are in phase when the power factor is unity. As a result, I_a is pulled onto the voltage phasor by itself, causing OB to equal Ia. Draw the voltage drop phasor caused by armature resistance using the voltage phasor's extreme point (point A). It is drawn so that AB = $-I_aR_a$, which is the opposite of the existing phasor. Point B is used to calculate the decrease caused by synchronous reactance. It is drawn so that BC = $-I_aX_s$ by being parallel to the IaRa phasor. Add AC to get the overall voltage loss. Join OC back emf per phase, where OC is the phasor E_b. The torque angle δ is the difference in phase between V and E_b. Thus,

$E_b = V - I_a R_a - jI_a X_s = V - I_a Z_s$

If equation (2) were taken into consideration while drawing the phasor diagram, it would resemble the image below.

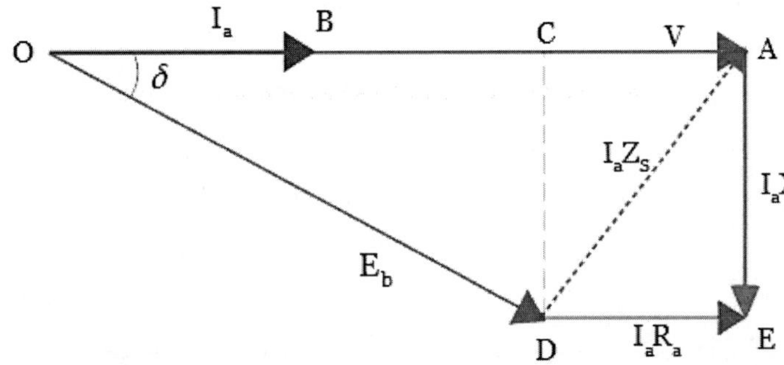

Fig.7.10 (b): Phasor diagram of Cylindrical Rotor Synchronous Motor for Unity Power Factor

(iii) Phasor Diagram for Leading Power Factor

Consider, however, that the synchronous motor is obtaining a current from the source at the leading power factor. The phasor diagram is once more drawn using the same equation (3).

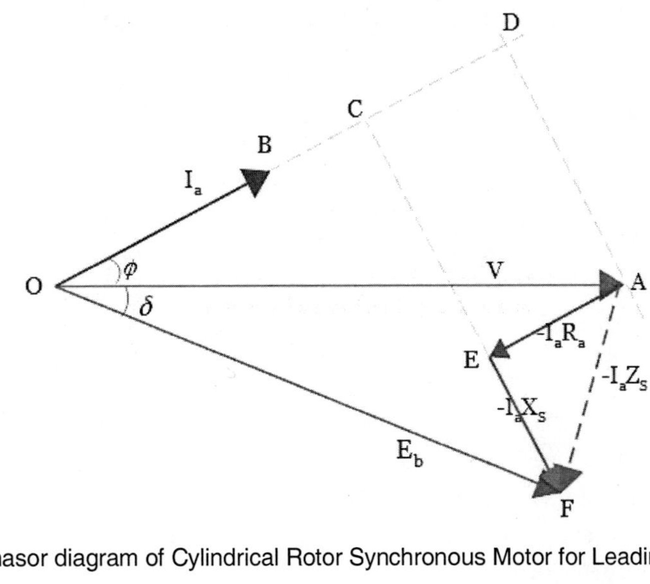

Fig.7.11 (a): Phasor diagram of Cylindrical Rotor Synchronous Motor for Leading Power Factor

Draw OA = V using the source voltage's phasor value as a reference. Since the armature current precedes the stator voltage, the phasor representing armature current I_a is drawn at an angle φ to the voltage phasor V. Draw the I_aR_a drop's phasor from point A in a manner that it is positioned in opposition to the one that is now present (AE = $-I_aR_a$). The IaXS drop is drawn perpendicular towards the I_aR_a phasor (EF = $-I_aX_s$) starting at the extremity of the phasor (from the point E). To get the overall voltage drop ($-I_aZ_s$), join AF. Join OF, which stands for the induced emf phasor per phase. The load angle is now determined by the phasor that exists between V and E_b.

Hence,

$$E_b = V - I_a R_a - jI_a X_s = V - I_a Z_s$$

You may create the phasor diagram for the leading power factor by using equation (2) as shown below.

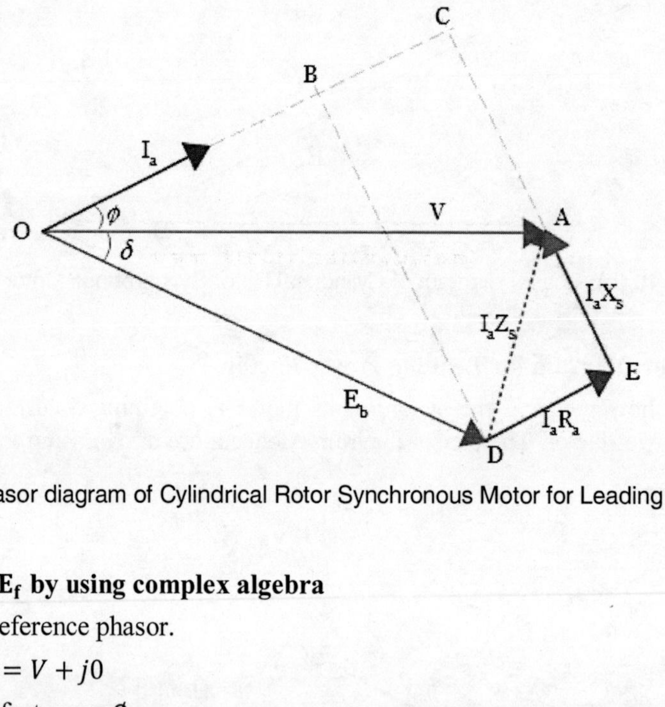

Fig.7.11 (b): Phasor diagram of Cylindrical Rotor Synchronous Motor for Leading Power Factor

Determination of E_f by using complex algebra

Let V be taken as reference phasor.

$$V = V\angle 0° = V + j0$$

For lagging power factor cos Ø

$$I_a = I_a\angle -Ø = I_a \cos Ø - jI_a \sin Ø$$

For unity power factor

$$I_a = I_a\angle 0° = I_a + j0$$

For leading power factor

$$I_a = I_a\angle +Ø = I_a \cos Ø + jI_a \sin Ø$$

The synchronous impedance is given by

$$Z_s = R_a + jX_s$$

The excitation voltage is given by

$$E_f = V - I_a Z_s$$

For lagging power factor $\cos \emptyset$

$$E_f \angle \delta = V \angle 0° - (I_a \angle -\emptyset)(R_a + jX_s)$$

$$= V + j0 - (I_a \cos \emptyset - jI_a \sin \emptyset)(R_a + jX_s)$$

$$= (V - I_a R_a \cos \emptyset - I_a X_s \sin \emptyset) - j(I_a X_s \cos \emptyset - I_a R_a \sin \emptyset)$$

$$E_f = \sqrt{(V - I_a R_a \cos \emptyset - I_a X_s \sin \emptyset)^2 + (I_a X_s \cos \emptyset - I_a R_a \sin \emptyset)^2} \quad \text{-------(7.4)}$$

$$\delta = \tan^{-1} \frac{I_a R_a \sin \emptyset - I_a X_s \cos \emptyset}{(V - I_a R_a \cos \emptyset - I_a X_s \sin \emptyset)}$$

Similarly for leading power factor $\cos \emptyset$

$$E_f = \sqrt{(V - I_a R_a \cos \emptyset + I_a X_s \sin \emptyset)^2 + (I_a X_s \cos \emptyset + I_a R_a \sin \emptyset)^2} \quad \text{-------(7.5)}$$

$$\text{and } \delta = -\tan^{-1} \frac{I_a R_a \sin \emptyset + I_a X_s \cos \emptyset}{(V - I_a R_a \cos \emptyset + I_a X_s \sin \emptyset)} \quad \text{-------(7.6)}$$

For unity power factor ($\cos \emptyset = 1$) $E_f = \sqrt{(V - I_a R_a)^2 + (I_a X_s)^2}$ -------(7.7)

$$\text{and } \delta = -\tan^{-1}\left(\frac{I_a X_s}{(V - I_a R_a)}\right) \quad \text{-------(7.8)}$$

Determination of E_f by phasor diagram

For lagging power factor shown in Fig.7.9(a)

From triangle ODM

$$OD^2 = OB^2 + BD^2 = OB^2 + CE^2 = (OC - BC)^2 + (CA - EA)^2$$

$$E_f^2 = (V \cos \emptyset - I_a R_a)^2 + (V \sin \emptyset - I_a X_s)^2 \quad \text{-------(7.9)}$$

For unity power factor load shown in Fig.7.10(a)

$$OD^2 = OC^2 + CD^2$$

$$E_f^2 = (V - I_a R_a)^2 + (I_a X_s)^2 \quad \text{-------(7.10)}$$

For leading power factor shown in Fig.7.11(a)

$$OD^2 = OB^2 + BD^2 = (OC - BC)^2 + (CA + AE)^2$$

$$E_f^2 = (V \cos \emptyset - I_a R_a)^2 + (V \sin \emptyset + I_a X_s)^2 \quad \text{-------(7.11)}$$

7.6 DIFFERENT TORQUES IN A SYNCHRONOUS MOTOR

Understanding the properties of a synchronous motor requires to understand the necessary torque to operate it at all times, from its beginning condition to its final shutdown. The varieties of torques connected with the synchronous motor need also be known in order to use it for various purposes.

The synchronous motor produces four forms of torque: pull-in torque, pull-out torque, beginning torque, and running torque. Let us go through them in depth.

- **Starting Torque**

Starting torque is additionally known as locked rotor torque. It is the minimal torque created by a synchronous machine with the rotor locked (i.e. stationary rotor) irrespective of the angular location of the rotor when rated voltage and frequency are applied. The motor's armature windings supply the locked rotor torque. It is also known as breakaway torque. It is the amount of torque necessary to rotate the motor from a standstill state. For centrifugal pumps, it can be as low as 10% and as high as "200 or 250% of full load torque" for loaded reversible of two-cylinder compressors.

- **Running Torque**

Running torque is the torque generated by a synchronous machine while it is running. The running torque is influenced by the motor's power rating and speed. The maximum operating torque needed by the machine being driven is determined by peak output power.

- **Pull-in Torque**

Because the synchronous motor isn't self-starting, it begins operating as an induction motor & operates at a speed that is 2% to 5% slower than the synchronous speed. The DC voltage for excitation is subsequently provided, and the rotor pulls through phase with the synchronously spinning magnetic field of the stator. Pull-in torque is the greatest amount of torque at rated frequencies and voltages that a synchronous motor will generate, when DC excitation is given to the motor.

- **Pull-out Torque**

The amount of load which can be provided to a synchronised motor has a limit. As the load increases, so does the torque angle or load angle () until a stage is achieved. Because the motor torque is less than the load torque at this point, the rotor becomes pulled out of synchronism & the motor comes to rest. As a result, pull-out torque, also known as breakdown torque, is the greatest amount of load torque which a synchronous motor can produce at its rated frequency and voltage without losing synchronism.

7.7 POWER FLOW EQUATIONS FOR A SYNCHRONOUS MOTOR

The equivalent circuit shown in Fig.7.8 has the phasor diagram at lagging power factor shown in Fig.7.9(a).

Here E_f lags behind V by angle δ so that

$$V = V\angle 0°, E_f = E_f\angle -\delta$$

By KVL

$$V = E_f + I_a Z_s$$

$$I_a = \frac{V-E_f}{Z_s} = \frac{V\angle 0°}{Z_s\angle\theta_z} - \frac{E_f\angle -\delta}{Z_s\angle\theta_z} = \frac{V}{Z_z}\angle\theta_z - \frac{E_f}{Z_z}\angle -(\delta+\theta_z)$$

$$I_a^* = \frac{V}{Z_s}\angle\theta_z - \frac{E_f}{Z_s}\angle(\delta + \theta_z)$$

Complex Power input to Motor per phase (S_{im})

$$S_{im} = P_{im} + jQ_{im} = VI_a^*$$

$$= \frac{V^2}{Z_s}\angle\theta_z - \frac{VE_f}{Z_s}\angle\delta + \theta_z$$

$$\therefore P_{im} + jQ_{im} = \left(\frac{V^2}{Z_s}\cos\theta_z + \frac{V^2}{Z_s}\sin\theta_z\right) - [\frac{VE_f}{Z_s}\cos(\delta + \theta_z) + j\frac{VE_f}{Z_s}\sin(\delta + \theta_z)] \qquad \text{-------(7.12)}$$

Real Input Power per Phase to the motor (P_{im})

Equating the real parts of Eq. (7.12) we get P_{im}

$$P_{im} = \frac{V^2}{Z_s}\cos\theta_z - \frac{VE_f}{Z_s}\cos(\delta + \theta_z)$$

$$P_{im} = \frac{V^2}{Z_s^{\,2}}R_a - \frac{VE_f}{Z_s}\cos(\delta + \theta_z) \qquad \text{-------(7.13)}$$

But $\theta_z = 90° - \alpha_Z$

$$\cos(\delta + \theta_z) = \cos\{90° + (\delta - \alpha_Z)\} = -\sin(\delta - \alpha_Z)$$

$$\therefore P_{im} = \frac{V^2}{Z_s^{\,2}}R_a + \frac{VE_f}{Z_s}\sin(\delta - \alpha_Z)$$

Reactive input Power per Phase to the motor (Q_{im})

Equating the imaginary parts of Eq. (7.12) we get Q_{im}

$$Q_{im} = \frac{V^2}{Z_s}\sin\theta_z - \frac{VE_f}{Z_s}\sin(\delta + \theta_z)$$

$$Q_{im} = \frac{V^2}{Z_s^{\,2}}X_s - \frac{VE_f}{Z_s}\sin(\delta + \theta_z) \qquad \text{-------(7.14)}$$

But $\theta_z = 90° - \alpha_Z$

$$\sin(\delta + \theta_z) = \sin\{90° + (\delta - \alpha_Z)\} = \cos(\delta - \alpha_Z)$$

$$Q_{im} = \frac{V^2}{Z_s^{\,2}}X_s - \frac{VE_f}{Z_s}\cos(\delta - \alpha_Z)$$

Complex power output per phase of the motor (S_{om})

$$S_{om} = P_{om} + jQ_{om} = E_f I_a^*$$

$$= E_f \angle -\delta \left(\frac{V}{Z_s} \angle \theta_z - \frac{E_f}{Z_s} \angle (\delta + \theta_z) \right)$$

$\therefore P_{om} + jQ_{om}$

$$= \frac{VE_f}{Z_s} \cos(\theta_z - \delta) + j\frac{VE_f}{Z_s} \sin(\theta_z - \delta) - \left\{ \frac{{E_f}^2}{Z_s} \cos\theta_z + j\frac{{E_f}^2}{Z_s} \sin(\theta_z) \right\} \text{-------(7.15)}$$

Real Power Output Per Phase of the Motor (P_{om})

Equating the real parts of Eq. (7.15) we get P_{om}

$$P_{om} = \frac{VE_f}{Z_s} \cos(\theta_z - \delta) - \frac{{E_f}^2}{Z_s} \cos\theta_z$$

$$P_{om} = \frac{VE_f}{Z_s} \cos(\theta_z - \delta) - \frac{{E_f}^2}{{Z_s}^2} R_a$$

But $\theta_z = 90° - \alpha_Z$

$$\cos(\theta_z - \delta) = \cos\{90° - (\delta + \alpha_Z)\} = \sin(\delta + \alpha_Z)$$

$$P_{om} = \frac{VE_f}{Z_s} \sin(\delta + \alpha_Z) - \frac{{E_f}^2}{{Z_s}^2} R_a \qquad \text{-------(7.16)}$$

Reactive Power Output Per Phase of the Motor (Q_{om})

Equating the imaginary parts of Eq. (7.15) we get Q_{om}

$$Q_{om} = \frac{VE_f}{Z_s} \sin(\theta_z - \delta) - \frac{{E_f}^2}{Z_s} \sin\theta_z$$

$$Q_{om} = \frac{VE_f}{Z_s} \sin(\theta_z - \delta) - \frac{{E_f}^2}{{Z_s}^2} X_s$$

But $\theta_z = 90° - \alpha_Z$

$$\sin(\theta_z - \delta) = \sin\{90° - (\delta + \alpha_Z)\} = \cos(\delta + \alpha_Z)$$

$$\therefore Q_{om} = \frac{VE_f}{Z_s} \cos(\delta + \alpha_Z) - \frac{{E_f}^2}{{Z_s}^2} X_s \qquad \text{-------(7.17)}$$

For a synchronous motor, power at the shaft = P_{om} – rotational losses

P_{om} is the mechanical power developed or gross power developed. Rotation losses include friction, windage and core losses.

Maximum Power Output of the Motor

For maximum power output of the motor

$$\frac{dP_{om}}{d\delta} = 0 \text{ and } \frac{d^2P_{om}}{d\delta^2} \angle 0$$

Differentiating equation 7.17w.r.t δ and equating it to zero.

$$\frac{d}{d\delta} = \left(\frac{VE_f}{Z_s}\sin(\delta + \alpha_Z) - \frac{E_f^{\ 2}}{Z_s^{\ 2}} R_a\right) = 0$$

$$\frac{VE_f}{Z_s}\cos(\delta + \alpha_Z) = 0$$

$$\therefore\ \delta + \alpha_Z = 90° \Rightarrow \delta = 90° - \alpha_Z = \theta_z$$

The maximum power output of the motor is given by

$$P_{om(max)} = \left(\frac{VE_f}{Z_s} - \frac{E_f^{\ 2}}{Z_s^{\ 2}} R_a\right) \qquad \text{-------(7.18)}$$

This occurs at $\delta = \theta_z$ which defines the limit of steady state stability. $P_{om(max)}$ is also called the maximum power developed.

7.8 EFFECT OF EXCITATION AT CONSTANT MECHANICAL LOAD

In a synchronous motor, the stator current (I_a) is calculated by dividing the voltage-phasor resultant (E_r) across V and E_b with the synchronous impedance Z_s. One of the most essential characteristics of a synchronous motor is the ability to change the field excitation, it may be configured to range from trailing to leading power factor. Considering a synchronous motor with an unvarying mechanical load and a fixed supply voltage. Because the mechanical load and speed are both constant, the power input provided by the motor (=3 V*I_a *cos ϕ) is likewise constant. This indicates that the in-phase element I_a cos ϕ taken from the supply will stay constant. When the field excitation is altered, the back e.m.f E_b likewise changes. For differing quantities of field excitation, this causes a change in the phase position of I_a with respect to V and the synchronous motor. It is worth noting that the extreme of present phasor I_a are on the straight line AB. As a result, the motor's power factor cos ϕ changes. The synchronous motor's phasor diagram is shown in Figure 7.12.

Now consider what occurs in a synchronous motor whenever the excitation voltage gets higher or lowered.

(i) Under excitation

The motor is considered to be under-excited whenever the field excitation is such that $E_b < V$. Under these conditions, the electrical current I_a trails behind V, causing the motor power factor to lag, as illustrated in Fig.7.12 (a). This is simple to understand. Since $E_b < V$, the net voltage E_r has fallen and is turning clockwise. Because the angle (= 90°) among E_r and I_a is steady, phasor I_a also rotates clockwise, implying that current I_a lags behind the source voltage. As a result, the motor has a trailing power factor.

(ii) Normal excitation

The motor is considered to be regularly excited when the field excitation is at a level where E_b = V. This is seen in Figure 7.12 (b). It is worth noting that rising excitation (i.e., increasing E_b) causes the phasor E_r and therefore I_a to rotate anti-clockwise, implying that the I_a phasor has moved closer to the phasor V. As a result, p.f. rises while remaining behind. Because the input power (=3 V*I_a *cos ϕ) remains constant, the stator current I_a has to decrease as the p.f. increases. Assume that the field excitation is raised until the electrical current I_a is in phase alongside the applied voltage V, resulting in a synchronous motor p.f. of unity. At unity p.f., the resulting E_r and, hence, I_a are minimal for a given load.

(iii) Over excitation

If the impact of field excitation causes E_b to exceed V, the motor is considered to be overexcited. As seen in Fig.7.12 (c), current I_a leads V and the electric motor's power factor leads. It should be noted that E_r and hence I_a continue to rotate anti-clockwise from the typical excitation position. As a result, I_a is in command of V. According to the preceding discussion, if the synchronous motor's coil is under-excited, it produces a trailing power factor.

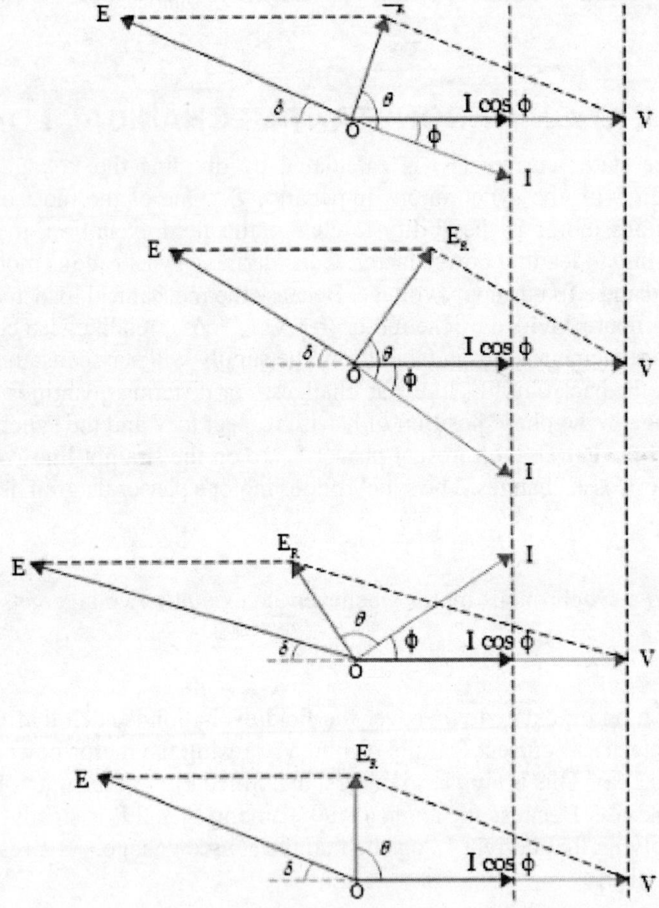

Fig.7.12 (a). Normal Excitation When E=V Lagging Power Factor

Fig.7.12 (b). Under Excitation When E<V More Lagging Power Factor

Fig.7.12 (c). Over Excitation When E>V Leading Power Factor

Fig.7.12 (d).Critical Excitation

Fig.7.12: Phasor diagram of Effect of Excitation at Constant Mechanical Load.

The power factor increases with increasing excitation until it reaches unity at usual excitation. Under these circumstances, the current taken from the supply of electricity is minimal. When the excitation is raised further (i.e., overexcited), the motor's power factor approaches leading. The current through the armature (I_a) is smallest at unity p.f. and grows as the electrical power factor deteriorates, either leading or trailing.

(iv)Critical excitation

The region of excitation is raised until the point when the angle of phase between voltage V and current I is zero. This type of excitation is known as critical excitation. It signifies that the current remains in phase with the voltage V, as seen in Fig.7.12 (d). With a power factor of unity, the current being drawn will be the smallest. Extending the excitation beyond that limit causes the synchronous motor to draw current with a higher power factor.

Comparison of various excitations

The below table compares the effect of different excitation for synchronous motor with a fixed load.

Table 7.1: Comparison of various excitations

Excitation	Comparison of E & V	Power Factor	Armature Current
Normal	E=V	Lagging	Increases
Under	E<V	More Lagging	Increases
Over	E>V	Leading	Increases
Critical	E>V	Unity	Minimum

7.9 V-CURVES AND INVERTED V-CURVES

According to the preceding explanation, if excitation is changed from extremely low (under excitation) to extremely high (over excitation), current I_a reduces, reaches minimal at unity p.f., and then grows again. However, in nature, the initial trailing stream becomes unity and eventually leads. This is shown in Figure 7.13.

Excitation can be enhanced by raising the field current going through the synchronous motor's field winding. If a graph of the motor's armature current (I_a) vs field current (I_f) is shown, its form appears like a V. When similar graphs are created under different load circumstances, we get a family of curves that all look like Vs. These curves are known as synchronous motor V-curves. Figure 7.13 (a) depicts them. In contrast, if the power factor (cos) is plotted against the field current (I_f), the graph has the form of an inverted V. Inverted V-curves for synchronous motors are generated by charting p.f. against I_f at various load situations. Figure 7.13(b) depicts these curves.

Typically, the manufacturer provides the synchronous machine's V-curves so that the user may identify the consequent operation under a specific set of conditions.

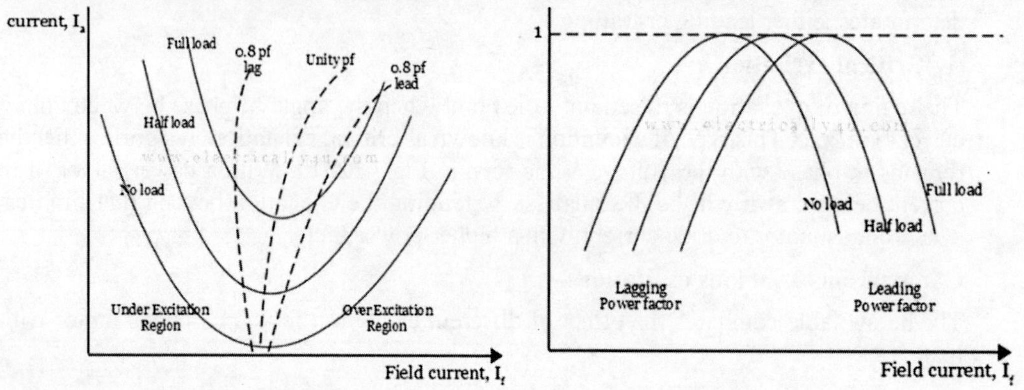

Fig.7.13: (a) **Fig.7.13:** (b)

Fig.7.13: V-Curves and Inverted V-Curves of synchronous motor

Experimental setup

V curves for various load circumstances may be obtained using a two-wattmeter approach setup with a synchronous motor. The synchronous motor is linked to the shunt wound direct current machine. The dc machine is initially started like a motor by supplying power for the field winding. When the motor reaches synchronous speed, the synchronous machine's field is activated. By then, the motor has achieved synchronism. When this is completed, the dc motor's power source is cut off, and it now serves like a load for the synchronous machine. Measure the ammeter, voltmeter, & wattmeter readings for different degrees of field excitation on no-load.

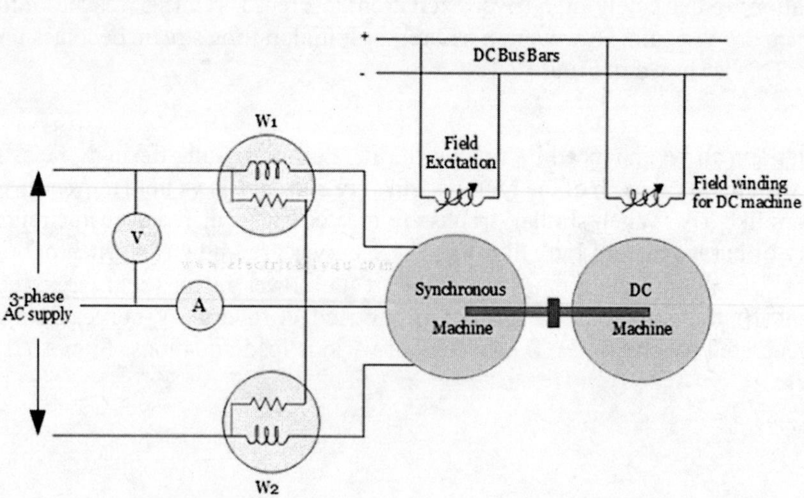

Fig.7.14: Experimental setup for V-Curves and Inverted V-Curves

The direct current (DC) generator is now loaded to the point where the motor is roughly half loaded. Analyse the values of the parameters within a metre for varying excitation while keeping the load constant. Rep the process for the whole load as well. The V shapes of the motor that is synchronous may be obtained by plotting the observed data on a graph.

7.10 HUNTING OR PHASE SWINGING

Hunting or Surging or Phase Swinging:

The load angle of an unloaded synchronous machine is zero degrees. As the shaft load increases, so does the load angle. Assume that load P1 is unexpectedly supplied to an unloaded machine shaft, causing the machine to slow down for a brief period. Also, the load angle (δ) grows from zero to one. During the initial swing, the electrical power produced equals the mechanical load P1. Because equilibrium has not been attained, the rotor swings further synchronous machine has zero-degree load angle. On increasing the shaft load gradually load angle will increase.

Let us consider that load P1 is applied suddenly to unloaded machine shaft so machine will slow down momentarily. The load angle reaches δ1 and becomes δ2. Load angle exceeds δ 1 and becomes δ2. The produced electrical power is now bigger than the prior one. The rotor reaches synchronous speed. However, it does not remain at synchronous speed and will continue to accelerate beyond synchronous speed. The load angle decreases as the rotor accelerates above synchronous speed. As a result, no balance is achieved once more. As a result, the rotor swings / oscillates around the new equilibrium point. This is referred to as hunting or phase swinging. Hunting happens not just within synchronous motors but additionally in synchronous generators when the load is abruptly changed.

Causes of Hunting in Synchronous Motor

- Sudden change in load.
- Sudden change in field current.
- A load containing harmonic torque.
- Fault in supply system.

Effects of Hunting in Synchronous Motor

- It may lead to loss of synchronism.
- Produces mechanical stresses in the rotor shaft.
- Increases machine losses and cause temperature rise.
- Cause greater surges in current and power flow.
- It increases possibility of resonance.

Reduction of Hunting in Synchronous Motor

Two techniques should be used to reduce hunting. These are

- **Use of Damper Winding:** It is made consisting of moderate electrical resistance copper or aluminium brushes that are placed in the slots of the pole faces of a salient pole machine. Damper winding reduces hunting by creating torque in the opposite direction of rotor slide. The degree of damping torque varies with slip speed.
- **Use of Flywheels:** A huge and hefty flywheel is provided with the prime mover. This improves the primary mover's inertia and aids in keeping the rotor speed fixed.
- Creating a synchronous machine with appropriate synchronising power coefficients
- Designing synchronous machine with suitable synchronizing power coefficients.

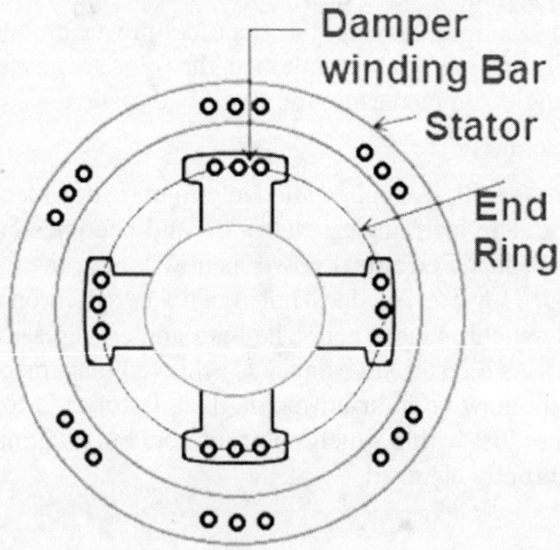

Fig.7.15 (a): Representation of Damper winding in Synchronous Motor

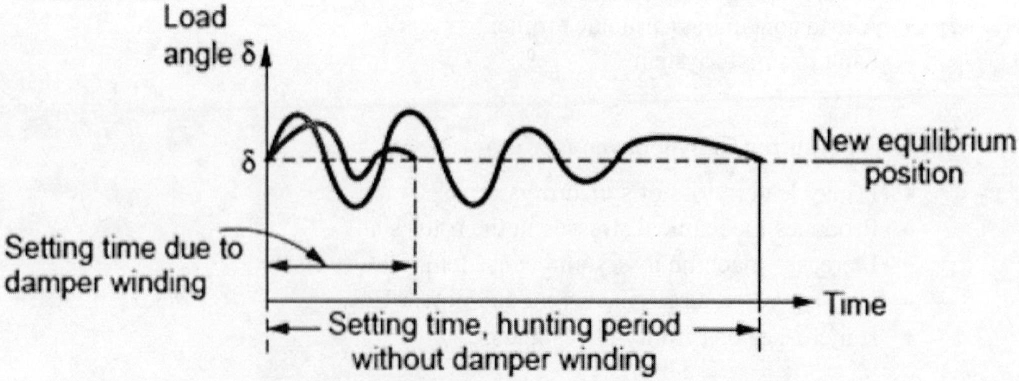

Fig.7.15 (b): Load angle vs. settling time

7.11 STARTING METHODS OF SYNCHRONOUS MOTORS

Because of the inertia of the rotor, synchronous motors are not naturally self-starting. When power is provided, the stator RMF immediately begins to rotate at synchronous speed. The rotor, on the other hand, cannot keep up. The following approaches are utilised to achieve the speed needed for the rotor properly to synchronise.

- **Damper Winding**

The salient pole rotor employs a damper winding. It is a short-circuited winding, similar to that of an induction motor. The RMF induces a current in this winding, which generates its own magnetic field, which interacts with the RMF and produces the necessary starting torque. When the rotor approaches synchronous speed, DC excitation is delivered to the rotor field winding, and the motor enters synchronism. The damper winding is used to start the motor using an induction motor in this way. This winding also aids in dampening oscillations caused by rapid changes in load.

- **Pony Motor Method**

Pony motors are tiny induction motors or direct current shunt motors that are linked to the shaft of synchronous motors. It contributes to the required starting torque. The direct current (DC) excitation is not used until the rotor has reached a speed close to synchronous speed. The rotor magnetically latches with RMF, and the pony motor's power source is turned off.

- **Variable Frequency Method**

A variable frequency drive, or VFD, is a device that produces electricity at a changeable frequency. As we all know, synchronous speed is determined by the supply frequency. To lower the synchronous speed, the frequency is first set to a minimum. The speed is progressively increased until it reaches the desired or regular pace.

7.12 LOSSES, POWER STAGES AND EFFICIENCY OF A SYNCHRONOUS MOTOR

A synchronous motor & an alternator (or else synchronous generator) are the same mechanism with distinct stages of power flow. When a machine is used to generate alternating current (AC) power (by converting input mechanical power towards output electrical power), this is referred to as a generator or synchronous generator. A synchronous motor is one that uses the same equipment to generate mechanical power (by turning input electrical energy through output mechanical energy). The losses in both machines are practically same, i.e. the losses with synchronous motor & synchronous generator (alternator) are almost identical except for the power flow stages.

Losses in Synchronous Motor

A motor's output mechanical power is always less than its input electrical power. The energy is lost as heat in various areas of the machine. It affects the motor's efficiency.

The synchronous motor losses are categorised into the following groups

- Copper Losses
- Iron and Core Losses
- Rotational and Frictional Losses

Copper Losses

Copper losses (which are also referred as electrical losses) are losses caused by current & winding resistance in copper windings. That is the reason, why it is often referred to as I^2R losses. It happens in both the stator and the rotor of a synchronous motor.

Stator Loss

It is caused by the input alternating current with stator winding resistance within the form of heat. It is provided

$$P_{stator} = (I_{stator})^2 R_{stator}$$

Where,

P_{stator} = power loss in stator

I_{stator} = input stator current (armature winding)

R_{stator} = resistance of stator (armature winding)

Rotor Loss

Because of the tiny DC rotor current passing through the field winding, the rotor loss is less than the stator loss. It is delivered by

$$P_{rotor} = (I_{rotor})^2 R_{rotor}$$

Where,

P_{rotor} = Power loss in rotor

I_{rotor} = DC rotor current (field winding)

R_{rotor} = Resistance of rotor (field winding)

Brushes Loss

Similarly the brushes loss is also considered copper I2R losses that account for rotor losses given by

$$P_{brush\ loss} = (I_{rotor})^2 R_{brush}$$

Iron and Core Losses

Magnetic Losses

The iron losses or core losses (also termed magnetic losses) are losses that take place in the synchronous motor's core or iron sections as a result of the material's magnetic characteristic. There are two types of losses: hysteresis losses and eddy current losses.

$$\text{Magnetic Losses} = P_{hys} + P_{eddy}$$

Hysteresis Loss

It develops as a result of the magnetization as well as demagnetization of the ferromagnetic core caused by a changing magnetic field. A ferromagnetic substance cannot change magnetization abruptly. During an electromagnetic reversal, it progressively demagnetizes as the supplied magnetic field quickly reverses. The demagnetization of the core requires a significant amount of applied power. This phenomenon is referred to as hysteresis loss and provided by

$$P_{hys} = \eta B_{max}^{1.6} fV$$

Where,

P_{hys} = Hysteresis losses

η = Hysteresis coefficient

B_{max} = Maximum flux density

f = Supply frequency

V = Volume of the magnetic material

Eddy Current Loss

Eddy current loss develops primarily as a result of the current produced in the synchronous motor's core. A fluctuating magnetic field, as we know, produces a current within a conductor, and the core is formed of iron, which is an excellent conductor. The induced current is known as Eddy current, and it flows in the core, squandering electric power in the form of heat, which is described as Eddy current loss. It is provided by

$$P_{eddy} = k_e B^2 f^2 t^2 V$$

Where,

P_{eddy} = Eddy current losses

K_e = Eddy current coefficient

B_{max} = Maximum flux density

f = Supply frequency

t = Thickness of lamination

V = Volume of the magnetic material

Laminating the core reduces Eddy current loss. To decrease the induced (Eddy) current, the core is made of thin sheets having lamination between them.

Rotational and Frictional Losses

Mechanical Losses

Rotation & friction losses (which is additionally referred as mechanical losses) occurs between the machine's stationary and moving elements. Since the rotor speed of a synchronous motor is constant, they are constant losses. A synchronous motor has two forms of mechanical losses.

Friction Losses

Friction losses in bearings occur in synchronous motors, owing to friction among the rotating and stationary parts. It is delivered by

$$P_{friction} = k\,N$$

Where 'k' is constant and 'N' is the speed in RPM

Windage Losses

Windage losses are caused by friction between spinning components and air. It is provided by

$$P_{windage} = k\,N^3$$

Windage losses rise with speed cube and are also affected by rotor design. Because of the protruding poles, a salient pole rotor offers larger windage losses. As a result, when designing the rotor, speed has to be taken into account.

Stray Losses

Stray losses are known as the difference among the overall losses and actual loss. These include a few minor losses, that caused in a synchronous motor for a variety of causes, but cannot be easily accounted, for example flux distortion, inconsistent current distribution in the armature, and so on. It is calculated as 1% or 0.01 from the overall losses.

$$P_{stray} = 1\%\ P_{total\_losses} = 0.01 \times P_{total\_losses}$$

Power Flow Diagram for Synchronous Motor

Power supplied in a synchronous motor is provided with a pin, and the mechanical power output is measured as P_{out}. P_{out} is the synchronous motor's real power output and the mechanical and electrical losses consume electricity and reduce the motor's real power output.

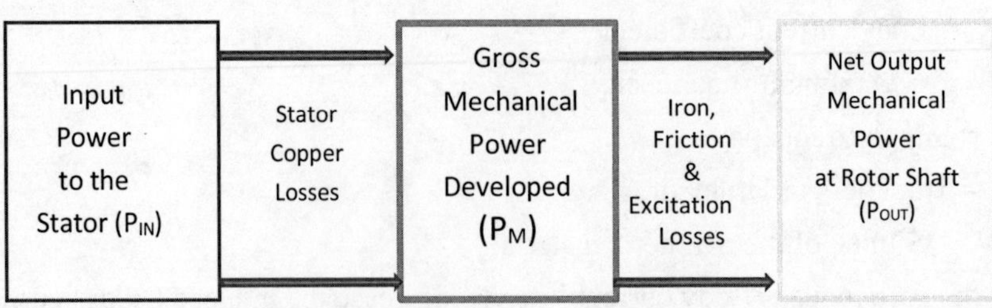

Fig.7.16: Power Flow Diagram for Synchronous Motor

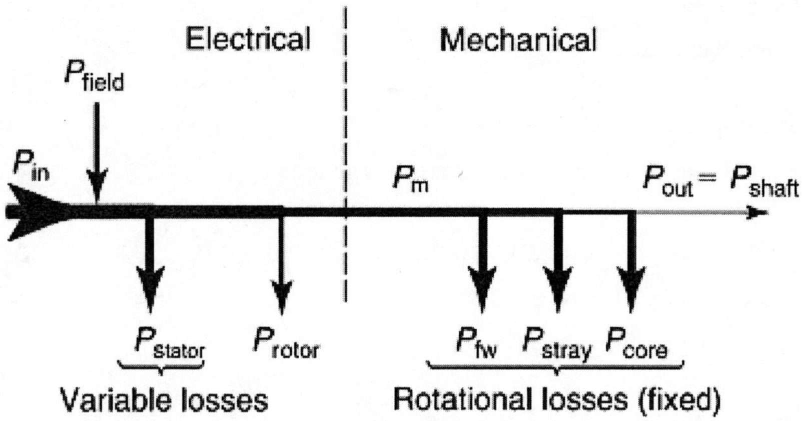

Fig.7.17: Power Stages of Synchronous Motor

Let's take a closer look at the phasor representation to better realise it. A synchronous motor's phasor schematic is shown below

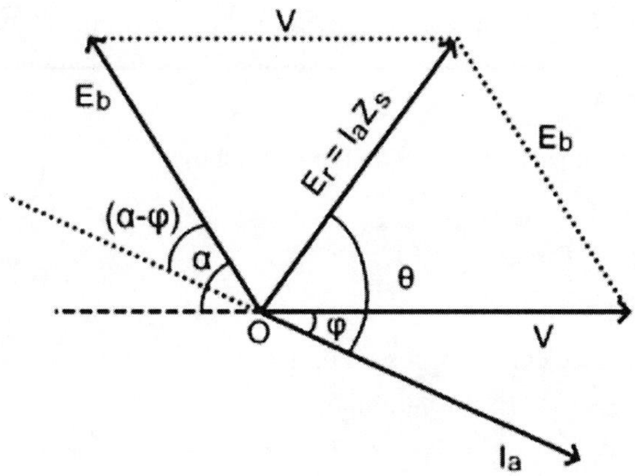

Fig.7.18: Phasor diagram for synchronous motor

Where

- V = Supply voltage per phase
- I_a = Armature current per phase
- R_a = Armature resistance per phase
- E_b = back emf
- E_r = resultant voltage, V – E_b
- Z_s = synchronous impedance

θ = internal angle, between I_a and E_r

α = Load angle

φ = Power factor angle

Input Electrical Power

The amount of input power for each phase is provided by

$$P_{ph} = VI_a \cos\varphi$$

The total three-phase input power required by a synchronous motor is provided by

$$P_{in} = \sqrt{3}\, V_L\, I_L \text{ Cos } \varphi$$

$$P_{in} = 3\, V_{ph}\, I_{ph} \text{ Cos } \varphi$$

Where

V_L = Line Voltage

I_L = Line Current

V_{ph} = Phase Voltage

I_{ph} = Phase Current

Gross Mechanical Power Developed

The power produced in the rotor is represented by the overall mechanical power Pm. It is computed by deducting the windings loss of the stator from the input power, as shown below.

$$P_{stator} = 3\, I_a^2\, R_a$$

Where 'I_a' and 'R_a' is the armature current and resistance.

$$P_m = P_{in} - P_{stator}$$

$$P_m = \sqrt{3}\, I_L V_L \text{ Cos } \varphi - 3\, I_a^2\, R_a$$

Or

P_m = back emf x armature current x cosine of the angle between them

$$P_m = E_b I_a \cos(\alpha - \varphi) \text{ for lagging p.f}$$

$$P_m = E_b I_a \cos(\alpha + \varphi) \text{ for leading p.f}$$

Mechanical Power Output

The real mechanical power given to the shaft, subsequently the load is referred to as output power and it is computed by deducting magnetic, mechanical, and stray losses for mechanical power.

$$P_{out} = P_m - P_{magnetic} - P_{Mechanical}$$

Synchronous Motor Efficiency

A synchronous motor's efficiency is defined as the proportion of its output mechanical power P_{out} to its input electrical power P_{in}. It is provided by

$$\text{Efficiency, } \eta = P_{out} / P_{in}$$

$$\text{Efficiency, } \eta = (P_{in} - P_{losses}) / P_{in}$$

$$\text{\% Efficiency, } \eta = (P_{in} - P_{losses}) / P_{in} \times 100\%$$

Losses affect a synchronous motor's efficiency. The magnetic losses are constant with respect to speed. Therefore, the connected load determines the efficiency. The most effective motor is a synchronous motor, which has a maximum efficiency of about 90%.

7.13 ADVANTAGE & DISADVANTAGES OF SYNCHRONOUS MOTOR

Advantages of synchronous motor are

- It has a constant operating speed.
- Speed depends on supply frequency and does not vary with any change in load.
- This motor operate in lagging, unity and leading power factor by changing the field excitation.
- This motor is used for power factor improvement and called as synchronous condenser.
- Higher efficiency above 90% as compared to the induction motor.
- Cost-effective at a lower speed than an induction motor.

Disadvantages of synchronous motor are

- It is not self-starting and require other means to provide near synchronous starting speed.
- This motor stalls if the load exceeds beyond breakdown limit.
- Requires an external DC source for its rotor field excitation.
- Unless the VFD (variable frequency drive) is utilized to change its supply frequency, the motor's speed cannot be changed.
- Hunting occurs with sudden application of load.
- Frequent maintenance requires due to slip rings and brushes.
- More complicated and costlier than induction motors.

Comparison of Induction and Synchronous Motor

Table 7.2: Induction vs. Synchronous Motor

Sl. No.	Induction Motor	Synchronous Motor
1.	It is a single-excited machine that does not require dc excitation.	It's a doubly-excited machine that needs both DC and AC power.

Sl. No.	Induction Motor	Synchronous Motor
2.	It possesses inherent starting torque	It needs outside resources to start.
3.	It never runs at synchronous speed and its speed drops as the load increases.	Operates at synchronous speed
4.	Speed control is possible	Speed control is not possible
5.	Supply only mechanical loads	Supply mechanical loads and can be used for improving system power factor
6.	Operates only at lagging P.F.	Operate for in all P.F.
7.	Maximum torque is proportional to square of the supply voltage.	Maximum torque is proportional to the supply voltage

7.14 APPLICATIONS OF SYNCHRONOUS MOTOR

Applications of synchronous motor are

- Constant Speed Application: Constant speed applications where the speed does not vary with increasing load.
- Power Factor Correction: The power factor of the electrical circuit can be varied by changing the excitation of the synchronous motor. It is called a synchronous condenser.
- Frequency Changer: Synchronous motor are used to run an alternator to supply having a different frequencies. Such a synchronous motor is known as a frequency changer.
- Voltage Regulation: This motor can act as a variable capacitor by varying its excitation. It can be used for voltage regulation by controlling the reactive power in power transmission line.
- General Applications: Grinders, pulp beaters, rock crushers, metal rolling mills, cement mills, rubber and textile mills, ball mills, steel mills, fans, blowers, centrifugal pumps, air compressors.

7.15 TYPES OF SYNCHRONOUS MOTORS BASED ON ROTOR MAGNETIZATION

According to the magnetization of the rotor, there are primarily two classifications for synchronous motors.

DC Excited Motor

A DC source is utilized to excite the rotor of a synchronous motor through a slip ring. Field winding on the rotor is magnetized to produce a steady magnetic field that interacts with the rotating magnetic field on the stator.

Non Excited Motor

A non-excited synchronous motor can produce a magnetic field without the need of outside excitation. It is made of a permanent magnet that produces its own field, either independently or with the stator field.

The Non Excited Motor can be classified into the following types

- Permanent Magnet Synchronous Motor
- Hysteresis Synchronous Motor
- Reluctance Synchronous Motor

UNIT SUMMARY

The aim of this chapter is to understand the construction and working principle of the synchronous motor. The synchronous motor has such a complex construction. The principle of operation of the motor conveys clearly why it is called synchronous motor. Also, discussed about the method of starting and various phenomena of the three-phase synchronous motor. The advantages, disadvantages and application also discussed in detail.

EXERCISES

1. A 75KW, 400V,4 poles, 3 φ star connected synchronous motor has (R+jXs) = 0.04 + j 0.04Ω/ph. Calculate full load 0.8 P.F. leading emf/phase and the gross mechanical power developed. Assume efficiency of 92.5%.

Solution:

Output=75 kW, P = 4, $Z_s = 0.04 + j\,0.4$, Cos Φ = 0.8 lead, η = 92.5 %

E =? P_m=?

Input = Output/ η=75/0.925=81.08 kW

$P_i = \sqrt{3}\, V_L I_L \cos\varphi$

$I_L = P_i/\sqrt{3}\, V_L \cos\varphi$

$I_L = \frac{81.08 \times 1000}{\sqrt{3} \times 400 \times 0.8} = 146.2$ A

$E^2 = (V\cos\Phi - IR)^2 + (V\sin\Phi + IX_s)^2$

$V_{Ph} = V_L/\sqrt{3} = \frac{400}{\sqrt{3}} = 230.9$ V/Ph

$E^2 = (230.9 \times 0.8 - 146.2 \times 0.4)^2 + (230.09 \times 0.6 + 146.2 \times 0.4)^2$

E = 265.2 V/Ph $\quad |E| > |V|$

$P_m = P_i^2 - 3I^2R = 81.08 \times 10^3 - 3 \times (146.2)^2 \times 0.04 = 78.51$ kW

2. A 400 volt, 3- φ ,star connected cylindrical rotor synchronous motor has R+ jX_s = 0.5 + j7.5 Ω/Ph for an e.m.f. of 1.25 times the terminal voltage. Find the maximum mechanical power developed by the machine.

Solution:

V_L = 400 V, R+jJX_s= 0.5 + j7.5, E = 1.25 V, P_{max} = ?

$V_{Ph} = V_L/\sqrt{3} = \frac{400}{\sqrt{3}} = 230.9$ V/Ph

E = 1.25 × 230.9 = 288.6V

$P_{max} = \frac{VE - E^2 \text{ Cos } \theta}{Zs} = \frac{[230.9 \times 288.6] - 288.6^2 \times Cos\ 86.1}{7.51^2}$ =811.8 W/Ph

$Z_s = \sqrt{0.5^2 + 7.5^2} = 7.51\ \Omega$

$\text{Sin}^{-1} = \frac{7.5}{7.51} = 87°$

$P_{max} = 3P_{max} = 3 \times 8118.8 = 24.35$kW

$= \frac{24350}{746} = 32.6$ HP

3. A 20 Pole, 40 HP,660 V, 60Hz, 3φ, star connected synchronous motor is operating at no load with generated e.m.f. is exactly equal to applied voltage. At no-load the rotor is retired by 0.5 mechanical degree from its synchronous position X_s= 10Ω, R=1 Ohm. Find power input per phase, armature copper loss per phase and armature current/Ph. Neglect mechanical losses.

Solution:

P = 20 V, V_L = 660 V, $\delta = 0.5\ Mech$,R=1 Ohm,X_s=10 Ohm, I_a=? and Loss=?

$P_i = P_m + Cu_{loss}$ (P_m=0)

$V = V_L/\sqrt{3} = 660/\sqrt{3} = 381$

$Z_s = R + jX_s = 1 + j10 = 10.05 \angle 84.3$

$Pi = \frac{V^2 Cos\theta - VECos\ (\theta+\delta)}{Z_s}$

$Pi = \frac{381^2 \times \cos 84.3 - 381^2 \times \cos(84.3+5)}{10.05}$ =1258W/Ph

P_i=Losses = 1258W/Ph

$I^2R = 1258$

$I = \sqrt{\frac{1258}{1}}$ =35.5 A

(**Note:** Mechanical Ψ=P/2 X θ= 20/2 X 0.5=5 Electrical, At no load V=E, P_i= Cu. Losses Losses, P_m=0)

4. A 50 Hz, 4 pole, 3φ ,star connected synchronous motor has X_s = 120 Ω/Ph, R=0, the excitation is such that open circuit voltage is 13.2 kV. Motor is connected to 11 kV, 50 Hz supply. What is the maximum load in the motor supply before losing synchronism? What is the corresponding torque line current and power factor?

Solution:

f= 50Hz, P = 4, X_s = 120 Ω/ph , R = 0, Open circuit voltage = emf = 13.2 V

Line voltage = 11000 V, P_{max} = ?, T = ?, I = ?, CosØ = ?

Operated at max , so δ= 90° = θ

$$P_{max} = \frac{VE - E^2 Cos\theta}{Z_s}$$

$$V = \frac{11 \times 10^3}{\sqrt{3}} = 6350 \text{ V}$$

$$E = \frac{13200}{\sqrt{3}} = 7621 \text{ V}$$

$$P_m = \frac{VE}{X_s} sin\delta = 403.27 \text{ kw/Ph}$$

$$P_{m\,(max)} = \frac{6350 \times 7621 - 7621^2 Cos 90°}{\sqrt{0^2 + 120^2}}$$

$P_{m\,(max)}$ = 403.2 kw/ph

$$\omega_s = \frac{2\pi N}{60} = \frac{2\pi \times 1500}{60} = 157 \text{ Rad/sec}$$

$$N_S = \frac{120 \times 50}{4} = 1500 \text{ rpm}$$

$$T = \frac{P_{mMax}}{W_s} = \frac{403.2 \times 10^3}{157} = 2568.15 \text{ N-m/ph}$$

Total power = 3T = 7704 N-m

Since max load δ = 90°, so right angle triangle

$$IZ = \sqrt{E^2 + V^2} = \sqrt{7621^2 + 6350^2} = 9919 \text{ V/ph}$$

$$I = \frac{I_z}{Z} = \frac{9919}{120} = 82.6 \text{ A}$$

If R = 0

Output = input – $I^2R(0)$

Input = output

$P_i = P_m$ = 403.2 kW

VICosØ = 403.3 kw

$$CosØ = \frac{403.2 \times 10^3}{6350 \times 82.6}$$

CosØ = 0.769 (leading)

5. A 3-∅ , star connected, 2000V, synchronous motor has $R + jX_s = 0.2+j2.2\ \Omega/ph$. The input is 800kW at rated voltage and emf is 2500 volts. Calculate the armature current and power factor.

Solution:

$$V = \frac{2000}{\sqrt{3}} = 1154.7\ V/ph$$

$$E = \frac{2500}{\sqrt{3}} = 1443.3\ V/ph$$

$$Z = \sqrt{0.2^2 + 2.2^2} = 2.2\Omega/ph$$

$$\theta = cos^{-1}\frac{0.2}{2.2} = 84.78$$

$$P_i = \frac{800\times10^3}{3} = 266.67\ w/ph$$

$$P_i = \frac{V^2 Cos\theta - VECos(\theta+\delta)}{Z_s}$$

$$= \frac{800\times10^3}{3} = \frac{(1154.7)^2 cos84.78 - 1154.7\times1443.3cos(84.78+\delta)}{2.2}$$

$$= \frac{800\times10^3}{3} - 1154.7^2 cos84.78 = -1154.7 \times 1443.3\cos(84.78+\delta)$$

$$\cos(84.78+\delta) = -0.2792$$

$$84.78 + \delta = 106.21$$

$$\delta = 21.4$$

$$(IZ)^2 = V^2 + E^2 - 2VE\ Cos\delta$$

$$= 1154.7^2 + 1443.3^2 - 2 \times 1154.7 \times 1443.3 \times cos\ 21.4$$

$$IZ = 559.54$$

$$I = 559.54/2.2 = 254.63A$$

$$Cos\ \emptyset = P_i/VI = \frac{800\times10^3}{3\times1154.7\times254.63} = 0.906$$

6. A 3-∅, star connected, 400 V, synchronous motor takes a input power of 5472 watts at rated voltage, R=0, $X_s = 10\Omega$, if its excitation is adjusted such that the e.m.f. is equal to rated voltage of 400 V. Calculate the load angle (δ), armature current and power factor.

Solution:

$$\delta = 20^\circ, X_s = 10\Omega, Cos\ \emptyset = 0.98$$

$$V = 400/\sqrt{3} = 230.94\ V/Ph$$

$$P_i = 5472/3 = 1824\ W/Ph$$

$$X_s = 10\Omega/Ph$$

$$P_i = \frac{VE}{X_s}\sin\delta$$

$$1824 = \frac{230.94^2}{10} Sin\delta$$

$\text{Sin}\delta = \frac{1824 \times 10}{230.94^2}$

$\delta = 20^\circ$

$IZ^2 = V^2 + E^2 - 2ve\cos\delta$

$= 2*230.94^2 - 2*230.94^2*\cos20 = 6432.78$

$I = \sqrt{\frac{6432.78}{X_s}}$

$\text{Cos}\,\emptyset = P_i/VI = \frac{1824}{230.94*8.02} = 0.98$

7. A 400 V, 50 Hz, 3-phase, 37.3 kW, star connected synchronous motor has a full efficiency of 88%. Synchronous impedance of the motor is (0.2+j 1.6) ohm/ph. If the excitation of motor is adjusted to give a leading p.f. of 0.9, calculate for full load (i) the induced e.m.f. and (ii) the total mechanical power developed.

Solution:

V_L=400V , 3ϕ , f=50HZ, star connected, $P_0 = 37.5kW$, η=0.88 , R=0.2 Ω/ph ,

X_s=1.6 Ω/ph , cosϕ= 0.9(lead)

$P_{in} = P_0/\eta = 37.5/0.88 = 42.61$ KW

$P_{in} = \surd V_L I_L \cos\phi$

$I_L = \frac{P_{in}}{\sqrt{3}\,V_L\cos\phi} = \frac{42.61\times10^3}{\sqrt{3}\times400\times0.9} = 68.33\text{A};\ I_{ph} = I_l = 68.33A$

$V_{ph} = \frac{400}{\sqrt{3}} = 231V$

cosϕ=0.9 , sinϕ=0.436

$E_{ph} = \sqrt{(\cos\phi - IR)^2 + (V\sin\phi + IX_s)^2}$

$= [(231\times 0.9 - 68.33 \times 0.2)^2 + (231\times 0.436 + 68.33 \times 1.6)^2]^{\frac{1}{2}} = 285.99\text{V}$

$E_L = \sqrt{3} \times E_{ph} = \sqrt{3} \times 285.99 = 495.35V$

$P_{in} = P_{in} - 3I^2R = (42.61 \times 10^3) - 3 \times (68.33^2) \times 0.2 = 39.808\text{k}W$

8. A 220V, Δ connected, 50 HZ synchronous motor has a synchronous reactance of 2.5Ω and negligible armature resistance its friction and windage losses are 1.0 kW and its core losses are 1.0 kW. The shaft is supplying a 22.5KW load .Find the load angle and induced e.m.f. in the stator winding is 255V

Solution:

$V_L = 220V\ , \Delta\ , f = 50HZ\ , X_s = 2.5\Omega\ , R = 0\ , W_{f\&w} = 1.5KW\ , W_c = 1KW\ ,$

$E_L = 255V\ , P_0 = 22.5KW\ , \delta = ?$

$V_L = 220V$

$E_L = 255V$

$Z_S = R + JXs = 2.5\angle 90^0 = 2.5\ \Omega/ph$

$P_m = P_o +$ Stray losses$= 22.5 + 1 + 1.5 = 25KW$

$P_m = [VE\cos(\theta - \delta) - E^2\cos\theta]/Z_s$

$= [VE\cos(90 - \delta) - E^2\cos 90]/Z_s$

$= VE\sin\delta\,/Z_s$

$$\sin\delta = P_m \times Z_s/VE = \frac{\frac{25\times10^3\times2.5}{3}}{220\times255} = 0.37136$$

$\delta = \sin^{-1}(0.3713) = 21.79^0$

9. A 400V, 6 pole, 3-ϕ, 50Hz, star connected synchronous motor has negligible armature resistance and 4Ω synchronous reactance per phase. It takes a current of 15A at unit P.F., when operating with certain field current. If the load torque is gradually increased until the line current is 60A. The field current is remaining unchanged. Find the gross torque developed and the new P.F.

Solution:

V=400V, P= 6, 3ϕ, f=50Hz, $, R = 0, X_s = 4\Omega/ph, I_L = 15A, \cos\phi = 1$

$I_L = 60A\,, T_{gross} = ?\ pf = ?$

$V_{ph} = \frac{400}{\sqrt{3}} = 231V, Z_s = 0 + j4 = 4\angle 90^0$, $Z_s = 4\Omega/ph, \theta = 90^0, \phi = 0\,, I_{ph} = I_l = 15A.$

$E_R = IZ_S = 15 \times 4 = 60V$

$$E_{ph} = \sqrt{V^2 + {E_R}^2} = \sqrt{231^2 + 60^2} = 238.66V$$

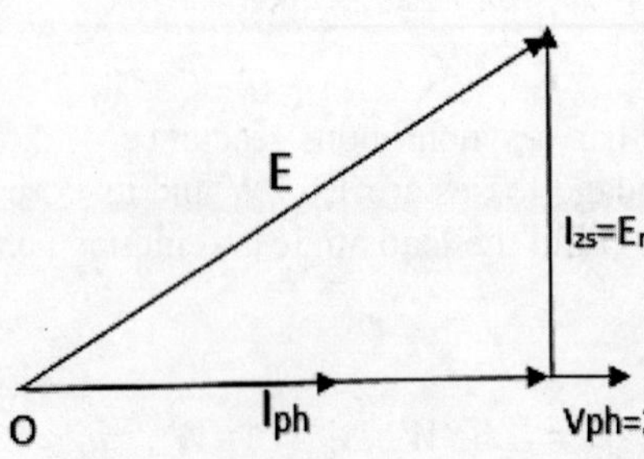

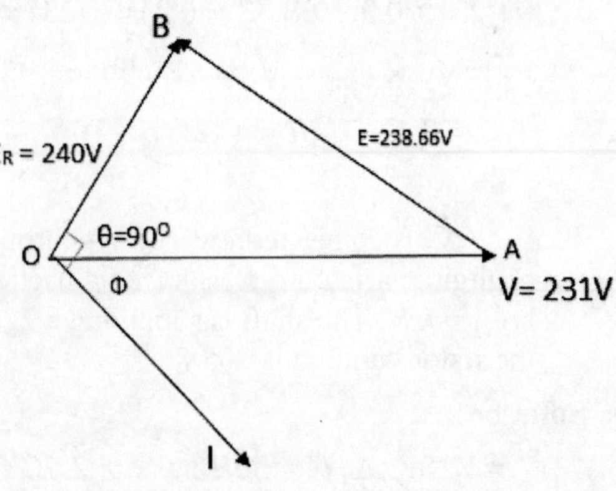

Case-2

When load torque gradually increases, the load angle increases. Hence the vector diagram is modified and it is given below.

$I_{ph} = I_l = 60A, V_{ph} = V = 231V\ , E_{ph} = 238.66V$

If $\alpha\ \phi_n \alpha E E_R = IZ_S = 60 \times 4 = 240V$

From triangle OAB

$\cos \angle AOB = (V^2 + {E_R}^2 - E^2)/2VE_R$

$= \frac{(231^2+240^2-238.66^2)}{2\times231\times240} = 0.487$

$(\cos\theta = \frac{a^2+b^2-c^2}{2ab})$

$\angle AOB = \cos^{-1} 0.487 = 60.86$

Now power factor angle $= \ \phi = \theta - \angle AOB = 90 - (60.86) = 29.14^2$

$p.f. = \cos(29.14) = 0.873(lag)$

$P_{in} = \sqrt{3} \times V_L \times I_L \cos\phi\ Watt$

$= \sqrt{3} \times 400 \times 60 \times 0.873$

$= 36289.9\ W = 36.28KW$

$P_{in} = P_{in} - 3 \times I^2R = 36.28 - 3 \times 60^2 \times 0 \times 10^{-3} = 36.28KW$

$$N_s = \frac{120f}{P} = \frac{120 \times 50}{6} = 1000rpm$$

$$P_n = T \times \omega = T \times \frac{2\pi N_s}{60}$$

$$T_g = \frac{P_m}{(\frac{2\pi N_s}{60})} = \frac{36.28 \times 10^3}{(\frac{2\pi\times1000}{60})} = 346\ N-m$$

10. A 500 V, 6 pole, 3Φ, 50 Hz, star connected synchronous motor has a resistance and synchronous impedance of 0.3Ω and 3Ω per phase respectively. The open circuit voltage is 600V. If the friction and core losses are 1kW, calculate the line current and power factor when motor output is 100 HP.

Solution:

Given are V_L = 500V, P = 6, 3Φ, f = 50Hz, Y, R= 0.3Ω/ph, X_s = 3Ω/ph, E = 600V, losses = 1kw, P_O= 100 HP. I_L and cosΦ=?

P_i = P_o + losses = 100 x 746+1= 75.6 kW

$P_i = \frac{3Ev\cos(\theta-\delta)}{Zs} - \frac{E^2}{Z_s}\cos\theta$

$V = \frac{500}{\sqrt{3}} = 288.7v,\ \theta = \tan\text{-}1\ \frac{Xs}{R} = \tan\text{-}1\ \frac{3}{0.3} = 84.3°$

$Z_s = \sqrt{R^2 + Xs^2} = \sqrt{(0.3)^2 + (3)^2} = 3.015\Omega$

$76.6 \times 10^3 = \frac{3 \times \frac{600}{\sqrt{3}} \times 288.7 \cos(84.3 - \delta)}{3.015} - \frac{3(\frac{600}{\sqrt{3}})^2 \cos 84.3°}{3.015}$

$\delta = 84.3 - 28.5 = 55.8°$

$(IZ)^2 = E^2 + V^2 - 2VE \cos\delta$

$= (346.4)^2 + (288.7)^2 - 2 \times 346.4 \times 288.7 \cos 55.8$

$(IZ) = 301.52$

$I = \frac{301.52}{Zs} = \frac{301.52}{3.015} = 100.008 \text{ A}$

(OR)

$I = \frac{V\angle 0 - E\angle -\delta}{Zs \angle \theta}$

$= \frac{288.7\angle 0 - 346.4\angle 55.8}{3.015\angle 84.3}$

$= 97.65 - j\,21.579$

$= 100.005\angle - 12.46°$

p.f = $\cos \Phi$

$= \cos 12.46$

$= 0.976$

Multiple Choice Questions

1. A 3- phase synchronous motor has
   a. High torque during starting
   b. No torque during starting
   c. Low current during starting
   d. Low torque during starting

2. A 3- phase synchronous motor needs dc supply for excitation
   a. Continuously
   b. At the starting instant only
   c. in stator
   d. None of these

3. Which of the following statements is/are incorrect for a synchronous motor?

   (i) It has starting torque.

   (ii) Its speed varies from no-load to full-load.

   (iii) It can operate at lt lagging, leading and zero power factors.

   a. (i),(ii) & (iii)
   b. Only (ii)
   c. Only (iii)
   d. & (ii)

4. Synchronous motors have

   a. Salient pole type rotor
   b. Smooth cylindrical type rotor
   c. Either salient type pole or smooth cylindrical type rotor
   d. None of the above

5. An ideal synchronous motor has no starting torque because the

   a. Rotor is made up of salient poles
   b. Relative velocity between the stator and the rotor m.m.fs. is zero
   c. Relative velocity between the stator and rotor m.m.fs. is not zero
   d. Rotor winding is highly reactive

6. In a synchronous motor

   a. Rotor m.m.f. and stator m.m.f. are stationary with respect to each other
   b. Rotor m.m.f. rotates somewhat faster in comparison to stator m.m.f.
   c. Stator m.m.f. rotates somewhat faster than rotor m.m.f.
   d. None of the above

7. A 3-phase synchronous motor speed

   a. Constant under all loads
   b. Varies with load
   c. Varies with half load
   d. Varies with power factor

8. A 3-phase synchronous motor

   a. The field m.m.f. leads the air gap flux and leads the armature m.m.f.
   b. The armature m.m.f. leads the air gap flux and air gap flux leads the field m.m.f.
   c. The armature m.m.f. leads the air gap flux and air gap flux lags behind the field m.m.f.
   d. None of the above

9. Armature current of a synchronous motor on no load

   a. Leads applied voltage by 90 degree.
   b. Lags behind applied voltage by 90 degree
   c. It is in phase will applied voltage
   d. Zero

10. As the load is increased on a synchronous motor, its speed

   a. Decreases
   b. Increases
   c. Remains constant and additional load is supplied by shift in relative position no of the rotor with respect to stator rotating magnetic field
   d. Remains constant for some time and then falls abruptly

11. Synchronous motor meets increase in load by taking more armature current as

   a. An increase in motor current results from the rotor pole moving backward relative to the stator pole.
   b. Motor current increases as the back e.m.f. drops.
   c. A stronger rotating field results in a rise in motor current.
   d. None of the above

12. Normal excitation is present in synchronous motor operation. The armature current drawn from the supply main increases with an increase in load because of

   a. Increase in back e.m.f.
   b. lowering of motor speed
   c. An increase in the voltage across the armature as a result.
   d. A rise in the power factor

13. The resultant voltage acting across the armature circuit of a synchronous motor (E_r) is the _of induced emf in the armature circuit (E_b) and supply voltage V

   a. Arithmetic sum
   b. Arithmetic difference
   c. Phasor difference
   d. Phasor sum

14. Phase e.m.f. induced in the armature of a synchronous motor (E_b) depends upon

   a. Rotor speed
   b. Load
   c. Both load and speed
   d. None of the above

15. In a synchronous motor, the synchronizing power come into action when

a. Rotor speed either exceeds or falls below the synchronous speed
b. Rotor speed is equal to synchronous speeds
c. Rotor speed falls below the synchronous speed
d. Rotor speed exceeds the synchronous speed

16.A synchronous motor's load angle, also known as the coupling angle, is the angle between

a. Rotor and stator poles of the same polarity
b. Rotor and stator poles of opposite polarity
c. Rotor and stator teeth
d. None of the above

17. The frequency, pole count, and load torque of a 3-phase, 7-pole, 50 Hz synchronous motor are halved. The motor will run at

a. 3000 r.p.m.
b. 1500 r.p.m.
c. 750 r.p.m.
d. None of the above

18. Exciters of synchronous machines are

a. D.C. shunt machines
b. D.C. series machines
c. D.C. compound machines
d. Any of the above

19. A synchronous motor is running on a load with normal excitation. Now if the load on the motor is increased

a. Power factor as well as armature current will decrease
b. Power factor as well as armature current will increase
c. Power factor will increase but armature current will decrease
d. Power factor will decrease and armature current will increase

20. The efficiency of a properly designed synchronous motor will usually fall in range

a. 60 to 70%
b. 75 to 80%
c. 85 to 95%
d. 99 to 99.5%

21. The p.f. of a synchronous motor is better than that of induction motor because

   a. Stator supply is releaved from producing the magnetic field.
   b. Mechanical load on the motor can be adjusted
   c. Synchronous motor runs at synchronous speed
   d. Synchronous motor has large air gap

22. In a synchronous motor, damper windings are provided on

   a. Stator frame
   b. Rotor shaft
   c. Pole faces
   d. None of the above

22. In a synchronous motor it the back e.m.f. generated in the armature at no-load is approximately equal to the applied voltage, then

   a. The motor is said to be fully loaded
   b. The torque generated is maximum
   c. The excitation is said to be zero percent
   d. The excitation is said to be hundred percent

24. Construction of a synchronous motor resembles with

   a. Series motor
   b. Induction motor
   c. Alternator
   d. Rotary converter

25. The breakdown torque in a synchronous motor is

   a. Proportional to the applied voltage
   b. Proportional to the square of the applied voltage
   c. Inversely proportional to applied voltage
   d. None of the above

26. A three-phase synchronous motor will have

   a. No slip-rings
   b. One slip-ring
   c. Two slip-rings
   d. Three slip-rings

27. Synchronous motor are operate at
    a. Lagging power factor only
    b. Leading power factor only
    c. Unity power factor only
    d. Any power factors

28. Synchronous motor working at leading power factor can be used as
    a. Voltage booster
    b. Phase advancer
    c. Noise generator
    d. Mechanical synchronizer

29. Which of the following motors the stator and rotor fields rotate simultaneously?
    a. D.C. motor
    b. Reluctance motor
    c. Universal motor
    d. Synchronous motor

30. Synchronous motor can be made self-starting by providing
    a. Damper winding on rotor poles face
    b. Damper winding on stator
    c. Damper winding on stator as well as rotor poles face
    d. None of the above

Answers of Multiple Choice Questions

1. b. No starting torque

2. a. Continuously

3. c. Only (iii)

4. a. Salient pole rotor

5. b. Relative velocity between the stator and the rotor mmfs is zero

6. a. The rotor mmf and stator mmf are stationary with respect to each other

7. a. Remains constant at all loads

8. b. The armature mmf leads the air gap flux and air gap flux leads the field mmf

9. a. Leads the applied voltage by 90 degree.

10. c. Remains constant and additional load is supplied by shift in relative position no of the rotor with respect to stator rotating magnetic field

11. a. The rotor pole falls back relative to the stator pole causing an increase in motor current

12. c. Increase in resultant voltage across the armature

13. c. Phasor difference

14. b. Load

15. a. Rotor speed either exceeds or falls below the synchronous speed

16. b. Rotor and stator poles of opposite polarity

17. b. 1500 r.p.m.

18.a. D.C. shunt machines

19.d. Power factor will decrease and armature current will increase

20. c. 85 to 95%

21. a. Stator supply is relieved of responsibility of producing magnetic field

22. c. Pole faces

23. d. The excitation is said to be hundred percent

24. c. An alternator

25. b. Directly proportional to applied voltage

26.c. Two slip-rings

27. d. Lagging, leading and unity power factors

28.b. Phase advancer

29. d. Synchronous motor

30.d. None of the above

Short and Long Answer Type Questions

Long Answer Questions with Answers

1. **Why 3Φ synchronous motor is not self-starting whereas 3Φ induction motor is self-starting?**

   In the case of synchronous motor stator winding is fed with 3Φ AC supply producing a RMF. The rotor poles are excited by a DC supply, hence poles are produced. It is only that when rotor poles and stator magnetic field are magnetically locked the rotor will continue to rotate at constant synchronous speed. If the rotor poles are not locked then rotor will come to rest and hence synchronous motor is not self-starting.

   In induction motor the rotor is not excited by a DC supply, but emf are induced in rotor conductor due to induction when 3Φ supply is fed to stator and hence a torque is produced making the motor self-starting.

2. Draw the families of curves of S.M. at no load and full load showing the relation between armature and field current. Also draw the P.F and field current using these curves explain how the motor 'overloaded' with no load connected to its shaft.

Families of curves of a S.M. at no load, half load and full load showing the relation between I_a and I_f and p.f and I_f are shown in the section 7.9respectively.

When the motor is over excited, the motor draws the leading current and shown by the current drawn by the motor minimum which is almost no-load current of the motor.

3. Explain what happened when the excitation of S.M is varies at constant load?

The change in the excitation of a synchronous motor operating on constant load causes the variation in the power factor of the motor when over excited synchronous motor has leading power factor shown in the section 7.8. However, when under excited, it has lagging p.f.

4. What are the methods of starting for synchronous motor?

Pony motor starting(using ac or dc motor)

- Starting as squirrel cage Induction motor or using damper winding
- Starting as slip ring Induction motor
- Using DC machine coupled to it.

5. Why damper windings are used in synchronous motor? Or How synchronous motor is started as squirrel cage motor?

Copper bars that have been short-circuited and are inserted into the rotor poles' face make up the damper winding. The stator winding of a three-phase synchronous motor produces a rotating magnetic field when an a.c. supply is applied to the stator. The damper winding, which is a part of the synchronous motor's rotor winding, induces e.m.f. and, as a result, currents begin to flow. Thus, torque affects the rotor. Synchronous motor damper windings serve the same purpose as induction motor rotor windings. So the synchronous motor starts as an induction motor and continues to accelerate because of the damper windings. The rotor windings are linked to exciter terminals when the motor reaches roughly 95% of the synchronous speed, and the rotor is magnetically locked by the rotating magnetic field.

6. What is the function of damper winding in synchronous motor?

In synchronous motor the damper winding are provided for starting torque and to prevent or minimize hunting.

7. What is hunting?

The sudden changes of load on synchronous motor set up oscillations in the rotor. Such oscillation of rotor about its new equilibrium position is called hunting.

8. What are the effects of hunting?

The effects of hunting in synchronous motor are Loses synchronism, Develops mechanical stress on rotor shaft, Increase machine losses and increase temperature of machine.

9. What is pull in torque?

When a motor is operating as an induction motor, the torque needed to draw it into synchronism is known as the pull-in torque.

10. What is pull out torque?

Pull out torque is the greatest torque that a motor can produce without pulling out of synchronism. Its value ranges from 1.5 to 3.5 times the torque at full load.

11. What is the effect on speed if the load is increased on a 3 phase synchronous motor?

From no load to maximum load, the motor operating at constant frequency bus bars maintains a consistent speed of operation.

12. Why a synchronous motor is called a constant speed motor?

In synchronous motors, load increases cause an increase in load angle while speed remains constant. As the load increases, the load angle also increases. The motor desynchronizes when the electrical load angle approaches 90 degrees. Because of this, the motor must revolve at synchronous speed to maintain synchronism.

13. State the main features of synchronous motor.

The main features of synchronous motor are

- The motor is not inherently self-starting
- The speed of operation is always in synchronous with the supply frequency irrespective of load conditions
- The motor is capable of operating at any power factor.

14. In what way synchronous motor is different from other motors?

The same principle governs both dc and ac motor operation. Magnetic locking between the magnetic fields of the stator and rotor is what allows synchronized motors to work.

15. A synchronous motor starts as usual but fails to develop its full torque. What could it be the reason?

The reasons for fails may be due to exciter voltage may be too low, field poles may be reversed, there may be either open-circuit or short-circuit in the field.

16. What potential causes could there be for a 3-phase synchronous motor not to start?

The following reasons may be the potential causes

- Voltage may be too low.
- Too much starting load.
- Open circuit in one phase or short circuit.
- Field excitation may be excessive

17. What do you mean by an 'infinite bus bar'? What are its specialties?

An infinite bus-bar is an ideal voltage source having zero internal impedance. Its terminal voltage remains constant irrespective of the load connected to it. Bus bars receiving power from a large grid can be treated as an infinite bus for all practical purpose.

Long answer questions:

1. Discuss any two methods of starting of synchronous motors.
2. Write a short note on the constant excitation circle.
3. Derive an expression for the power developed per phase of a synchronous motor.
4. Explain the variation of current and power factor of a synchronous motor with excitation.
5. Derive an expression for torque developed in a synchronous motor.
6. Write short notes on the following:
   i)causes of Hunting and its suppression
   ii) Mathematical analysis of power developed in Synchronous motor
7. Explain the different methods of starting of synchronous motors.
8. Show that the locus of stator current for a constant output of 3-phase synchronous motor connected to a constant voltage, constant frequency bus-bars is a circle?
9. Explain the construction of damper winding. Clearly show the location of damper winding.
10. Derive the expression for the maximum power developed by a synchronous motor.
11. Draw and explain the phasor diagram of 3-phase synchronous motor when (i) it is over excited (ii) it is under excited.
12. What is meant by excitation circle? Explain construction of excitation circle for a synchronous motor.
13. Explain the operation and characteristics of synchronous induction motor.
14. Draw the phasor diagrams of synchronous motor for unity, leading and lagging power factor conditions.
15. List and Explain the Starting methods for synchronous motor.
16. What are the merits and demerits of synchronous motor?
17. Explain the procedure to determine the V and inverted V- curves of a synchronous motor.
18. Discuss the principle of operation of a synchronous motor. Also list their applications.
19. Derive an expression for power developed in a cylindrical rotor synchronous motor in terms of load angle and synchronous impedance.
20. Explain the construction of 'excitation circle' for a synchronous motor.

Short answer Questions:

1. What are various applications of synchronous motors?
2. What could be the reasons if a 3-phase synchronous motor fails to start?
3. What is meant by hunting in synchronous motors and how it can be avoided?
4. What is the function of synchronous condenser?
5. What is power circle of a synchronous motor?
6. Explain the principle of operation of Synchronous motor.
7. What is a synchronous condenser? What are its applications?
8. Why a 3-phase synchronous motor will always run at synchronous speed?
9. How the starting current limited in case of a synchronous motor provided with a damper winding?
10. What are the effects of increase in excitation of a synchronous motor?
11. How is the hunting avoided in a synchronous motor?
12. Draw the phasor diagram of a salient pole synchronous motor.
13. How can the speed of the synchronous motor varied?
14. What is synchronous condenser? List any two applications of it.
15. What is hunting phenomenon?

Numerical Problems

1. A synchronous machine has a synchronous reactance of 1.0 ohm per phase and an armature resistance of 0.1 ohm per phase. If $E_A=460\angle-10^0$V and phase voltage is $480\angle-0^0$V. Is the machine a motor or generator? How much power P is this machine consuming from or supplying to the electrical system? How much reactive power Q is this machine consuming from or supplying to the electrical system?

[**Ans.** This machine is a motor, It is consuming reactive power, P=125.3 kW and Q=28.9 kVAR.]

2. A 400 V, three phase, star connected synchronous motor takes 3.73 kW at normal voltage and has an impedance of (1+j8) ohm per phase. Calculate the current and p.f. if the induced e.m.f. is 460V.

[**Ans.** 6.28 A, 0.86 lead]

3. A 2200 V, three phase, star connected synchronous motor has a resistance of 0.6 ohm and a synchronous reactance of 6 ohm. Find the generated e.m.f. and the angular retardation of the motor when the input is 200 kW at power factor unity and power factor 0.8 leading.

[**Ans.** 2.21 kV, 14.3^0 and 2.62 kV, 12.8^0]

4. The synchronous reactance per phase of a three phase star connected 6600 V synchronous motor is 20 ohms. For a certain load, the input is 915kW at normal voltage and the induced line e.m.f. is 8942 V. Calculate the line current and the p.f. (Neglect the resistance).

[**Ans.** 97A and 0.8258 Leading]

5. The input to a 11kV, three phase, and star connected synchronous motor is 60 A. The effective resistance and synchronous reactance per phase are 1 ohm and 30 ohms respectively. Find the power supplied to the motor and the induced e.m.f. for the p.f. of 0.8 leading.

[**Ans.** 915 kW and 13 kV]

6. A 10 kW, 400 volt, 3-phase, star connected synchronous motor has a synchronous impedance of 0.3 + j 2.5. Find the voltage to which the motor must be excited to give a full load output at 0.866 leading pf. Assume the armature efficiency of 90%. Also calculate the total mechanical power developed, losses in armature winding and iron plus excitation cases.

[**Ans.** E_b=252.96 Volts]

7. The line current of 11 kV, three phase, star connected synchronous motor is 60 amperes at a power factor of 0.8 leading. The effective resistance and synchronous reactance per phase are respectively 1 ohm and 30 ohms. Find the phase and the line value of the induced emf and also the power input of the motor.

[**Ans.** E_{ph}= 7528.92 angle − 11.3° Volts. E_L =13040.47 Volts]

8. The load in a factory is 800 kVA at 0.85 power factor lagging and in addition there is a synchronous motor having an input of 200 KW. Determine the input to the synchronous motor in kVA and the power factor at which it must operate, if the power factor of the combined load should be 0.95 lagging.

[**Ans.** KVA rating of the motor = 239.737 kVA, Power factor of the motor= 0 834 (leading)]

9. The full load current of a 6600 volt, 3-phase star connected synchronous motor is 80 amperes at 0.83 power factor leading. The resistance and synchronous reactance of the motor are 3 ohms and 22 ohms per phase respectively. Calculate the phase and line value of emf induced, the efficiency and the bhp of the motor. The stray losses are equal to 35 kW.

[**Ans.** Phase value of induced emf =4861 Volts, Line value of induced emf=8419.5 Volts, Efficiency=87.8 %, BHP of the motor= 893 37. BHP]

10. A 208 V, star connected, synchronous motor is drawing 50 A at unity power factor from a 208 V power system. The field current flowing under these condition is 2.7 A. Its synchronous reactance is 1.6 ohm. Assume a linear open circuit characteristic. Find the phase voltage of the motor, internal generated voltage and torque angle.

[**Ans.** phase voltage of the motor=120 $\angle 0^0$ V, internal generated voltage=144$\angle$-33.7^0 V and torque angle and torque angle= -33.7^0]

PRACTICAL

1. Identify various parts of the 3-phase synchronous motor and list various materials used for construction.
2. Identify various applications of synchronous motor and note the specifications of the motor in various applications.
3. Study the effect of variation of field current upon the stator current and power factor of a synchronous motor at various load and draw V-curves and invert V-curves.
4. Study how to operate the synchronous motor as a condenser?
5. Perform the speed control of synchronous motor as a modern drive.

KNOW MORE

1. Working of synchronous motor
   https://www.youtube.com/watch?v=Vk2jDXxZlhs
2. Construction & Working Principle of Synchronous Motor
   https://www.youtube.com/watch?v=JC73hjt6Plw
3. Understanding electric motor Windings!
   https://www.youtube.com/watch?v=YYQayMrK4Fo
4. Synchronous Motor
   https://www.youtube.com/watch?v=fdMIuEqh48M&list=PLPpCFgQP7QKHSJQnSwaigL89gshecycXs
5. Electrical Motors Manufacturing Process
   https://www.youtube.com/watch?v=9OpwrbdyMvU

REFERENCES AND SUGGESTED READINGS

1. Ashfaq Hussain, Haroon Ashfaq (2005), Electric Machines, Dhanpath Rai & Co. (P) Ltd., New Delhi(ISBN: 978- 81- 7700- 166-2)
2. P.S. Bimbhra, Electric Machines, Khanna Book Publishing Co., New Delhi (ISBN: 978- 93-6173- 294)
3. Mittle, V.N. and Mittle, Arvind., Basic Electrical Engineering, McGraw Hill Education New Delhi, ISBN :9780070593572
4. Kothari, D. P. and Nagrath, I. J., Electrical Machines, McGraw Hill Education. New Delhi, ISBN:9780070699670
5. Bhattacharya, S. K., Electrical Machines, McGraw Hill Education, New Delhi, ISBN:9789332902855
6. Theraja, B.L., Electrical Technology Vol-II (AC and DC machines), S.Chand and Co. Ltd., New Delhi, ISBN : 9788121924375
7. Sen, S. K., Special Purpose Electrical Machines, Khanna Publishers, New Delhi, ISBN: 9788174091529
8. Janardanan E. G, Special Electrical Machines, Prentice Hall India, New Delhi ISBN: 9788120348806
9. Hughes E., Electrical Technology, ELBS
10. Cotton H., Electrical Technology, ELBS

Dynamic QR Code for Further Reading

8 Fractional Horsepower (FHP) Motors - Working and Construction

UNIT SPECIFICS

Through this module we have discussed the following aspects:

- *Familiarizing the construction, working of Synchronous Reluctance Motor,*
- *Familiarizing the construction, working of Switched Reluctance Motor,*
- *Familiarizing the construction, working of BLDC motor,*
- *Familiarizing the construction, working of Permanent Magnet Synchronous Motors,*
- *Familiarizing the construction, working of Stepper motors,*
- *Familiarizing the construction, working of AC and DC servomotors,*
- *Torque speed characteristics of above motors,*
- *Applications of the above motors.*

The photos of practical motor are shown to detail the various parts of the machine. The real time applications of the topics are discussed for inducing further curiosity and creativity.

Besides giving a large number of multiple-choice questions as well as questions of short and long answer types marked in two categories following lower and higher order of Bloom's taxonomy, a list of references and suggested readings are given in the unit so that one can go through them for practice. It is important to note that for getting more information on various topics of interest some QR codes have been provided in different sections which can be scanned for relevant supportive knowledge.

After the related practical, based on the content, there is a "Know More" section. This section has been carefully designed so that the supplementary information provided in this part becomes beneficial for the users of the book. This section mainly highlights the initial activity, examples of some interesting facts, analogy, history of the development of the subject focusing the salient observations and finding, timelines starting from the development of the concerned topics up to the recent time, applications of the subject matter for our day-to-day real life or/and industrial applications on variety of aspects, case study related to environmental, sustainability, social and ethical issues whichever applicable, and finally inquisitiveness and curiosity topics of the unit.

RATIONALE

Special machines or Fractional horse power motors are the work horse of any industry. It has a widespread application in industrial and domestic applications. This chapter introduces the students to the construction, working principle and applications of the special machine. To understand the fundamentals in the principle of operation of these machines is very important aspect. The use of power electronics technology plays an important role for special machines.

PRE-REQUISITES

Machines I: Sem III

Physics: Electromagnetism (Class X)

UNIT OUTCOMES

List of outcomes of this unit is as follows:

U8-O1: Understand construction of the fractional horse power (FHP) motor

U8-O2: Able to understand the constructional differences in FHP motor

U8-O3: Feel the working principle of FHP motor

U8-O4: Why self-starting is not possible in FHP motor

U8-O5: Able to get to know about the applications of FHP motor

Unit-8 Outcomes	EXPECTED MAPPING WITH COURSE OUTCOMES (1- Weak Correlation; 2- Medium correlation; 3- Strong Correlation)				
	CO-1	CO-2	CO-3	CO-4	CO-5
U8-O1	-	-	-	-	3
U8-O2	-	-	-	-	3
U8-O3	-	-	-	-	3
U8-O4	-	-	-	-	3
U8-O8	-	-	-	-	3

8.1 INTRODUCTION TO SYNCHRONOUS RELUCTANCE MOTOR

A ferromagnetic rotor, like those seen in reluctance motors, induces non-permanent magnetic poles despite the absence of windings. This rotor uses magnetic reluctance to produce torque. This type of motor is one that is individually activated, and the rotor it uses is not symmetrical. Different types of reluctance motors are available, including synchronous, variable, switching, and variable stepping reluctance motors. Due to the difficulty in constructing and operating them, this motor had a limited use at the beginning of the twenty-first century. Therefore, this can be avoided by developing design tools, theory, and embedded systems.

One type of synchronous electric motor is the synchronous reluctance motor without field windings or permanent magnets and generates torque due to the difference in magnetic conductivities along the direct and quadrature axes of the rotor. Due to its simple & robust construction, this type of motor is currently becoming increasingly popular as a choice for hybrid and electric vehicles. The primary advantage of this motor is the absence of rotor cage losses, which allows for a permanent torque that is larger than that of an IM (Induction Motor) of the same size. The structure of Synchronous reluctance motor is shown in Fig.8.1.

Fig.8.1: Synchronous Reluctance Motor

The following are the major characteristics of synchronous reluctance motors.

- In comparison to induction motor drives, the field-based control algorithm is simplified.
- The precise torque can be chosen so that it has no impact on the rotor's temperature.
- This motor's rotor is less expensive than those of induction and permanent magnetmotors.

8.1.1 Construction of Synchronous Reluctance Motor

This motor's design is comparable to that of the salient pole synchronous motor. This motor's stator contains three phase symmetrical winding, but the rotor without any field winding. This winding will produce a sinusoidal rotating magnetic field within the air gap, and the induced magnetic field within the rotor can result in the development of a reluctance torque. This rotor has a tendency to link to the stator field with the least amount of reluctance possible. Using iron laminations in the axial direction that are divided by non-magnetic material, the rotor of the current reluctance motor can be designed. This motor's performance is comparable to that of an induction machine, but because it is simpler, less expensive, and has a sturdy structure, it may have a higher efficiency than an induction motor.

This motor is primarily intended for use in situations requiring high power. This can be categorised in the manners depicted in Fig.8.2 (a), (b), and (c).

- Salient pole rotor
- Axially Laminated Rotor,
- Transverse, or radial laminates

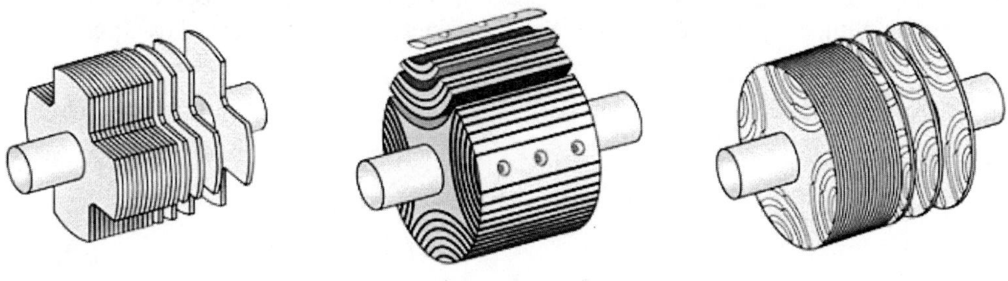

(a) (b) (c)

Fig.8.2: (a) Rotor with salient poles, (b) Axially laminated rotor and (c) Transversally laminated rotor

Reluctance motors are ideal for a variety of applications because they may give extraordinarily high power density at a lower cost. The main demerits are strong torque ripple when used at lower speeds, which also produces noise. This motor's stator mostly consists of prominent electromagnet poles that are comparable to those in a BLDC motor. A soft magnetic material, such as laminated Si steel, is used in the rotor. Several projections of this material function like prominent magnetic poles with magnetic resistance. The rotor poles of the motor have less than the stator poles, which minimises torque ripple and prevents the poles from connecting totally. A rotor pole is fully in the unaligned position when it is located in the midst of two adjacent stator poles. This is the highest point of magnetic reluctance for the rotor pole. Many rotor poles are totally attached to many stator poles and are in a less reluctant position when they are aligned.

Once a stator pole is turned on, the rotor's torque will change in a way that reduces reluctance. In order to connect through the stator field, the adjacent rotor pole can be withdrawn from its unconnected state. By consistently pulling the rotor, the stator field should rotate before the rotor poles in order to sustain revolution.

Certain motor types' substitutes will run on three-phase AC electricity. The majority of the designs that are currently on the market are of the switched reluctance type because electronic commutation offers significant control advantages for smooth operation, motor speed control, and starting. High efficiency at synchronous speed without the use of rare earth permanent magnets is one of synchronous reluctance motor's key characteristics. The distributed sinusoidal AC stator windings that are connected via a specific rotor lamination design are primarily supported by these motors. These motors' simple, low-inertia revolving assembly design enables synchronous speed operation. Fewer torque applications require for efficient operation can use synchronous reluctance motors.

8.1.2 Working Principle of Synchronous Reluctance Motor

A rotating magnetic field is produced in the electric motor's air gap by the alternating current entering through the stator windings. The Magnetic field lines of a synchronous reluctance motor is shown in Fig.8.3. In order to reduce the reluctance (magnetic resistance) in the magnetic circuit, the rotor tries to align itself with its most magnetically conductive axis (d-axis) using an applied field. The difference between the direct L_d and quadrature L_q inductances directly relates to the torque's amplitude. Therefore, the torque produced increases with increasing difference.

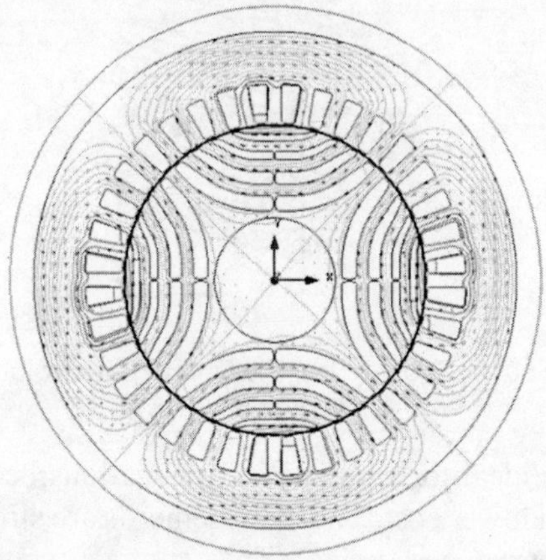

Fig.8.3: Magnetic field lines of a synchronous reluctance motor

The Fig.8.4 below can be used to explain the basic concept. The conductivity of the object "a" made of anisotropic material varies along the d and q axes, but the conductivity of the object "b" made of isotropic magnetic material is constant in all directions. If there is an angle between the lines of the magnetic field and the d axis, the torque produced by the magnetic field applied to the anisotropic object "a" will increase. The magnetic field will undoubtedly become distorted if the object's "a" d axis does not align with the lines of the magnetic field. In this instance, the deformed magnetic lines' direction will match the object's q axis.

A magnetic field is produced in a synchronous reluctance motor by a stator winding that is sinusoidally dispersed. The field can be thought of as sinusoidal because it rotates at synchronous speed. In such a case, there will always be a torque designed to lower the potential energy of the entire system by lowering the field distortion along the q axis ($\delta \rightarrow 0$).

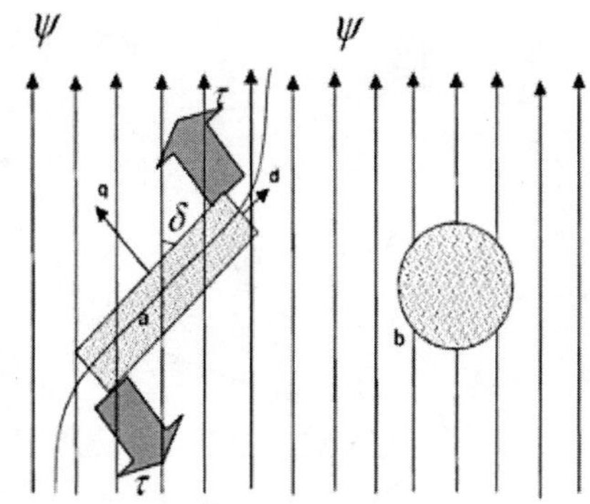

Fig.8.4 An object with (a) anisotropic geometry and (b) isotropic geometry in a magnetic field

The electromagnetic energy will continue to be transformed into mechanical energy if the angle is maintained, for instance by adjusting the magnetic field. The magnetization and the torque that the stator current produces aim to decrease field distortion. The current angle, or the angle between the current vector of the stator winding and the rotor d-axis in a rotating coordinate system, is used to adjust the torque.

8.1.3 (A) Phasor Diagram

Figure 8.5 below shows the synchronous reluctance motor's phasor diagram. A synchronous reluctance motor does not employ DC excitation. V_f is therefore excluded. Stator resistance is neglected. Ψ_s is the stator flux linkage.

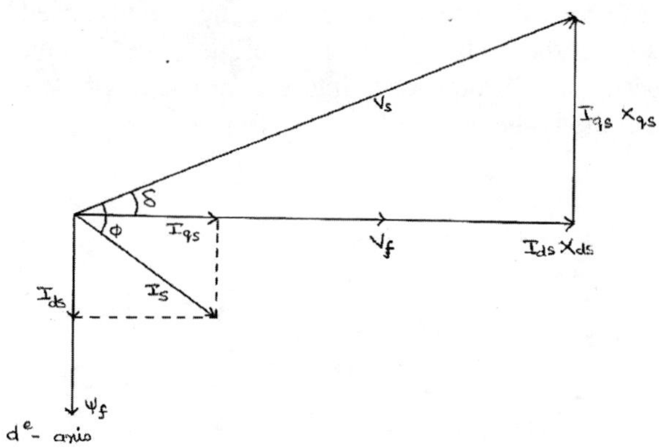

Fig.8.5 (a): Phasor diagram of the synchronous reluctance motor with q-axis as reference

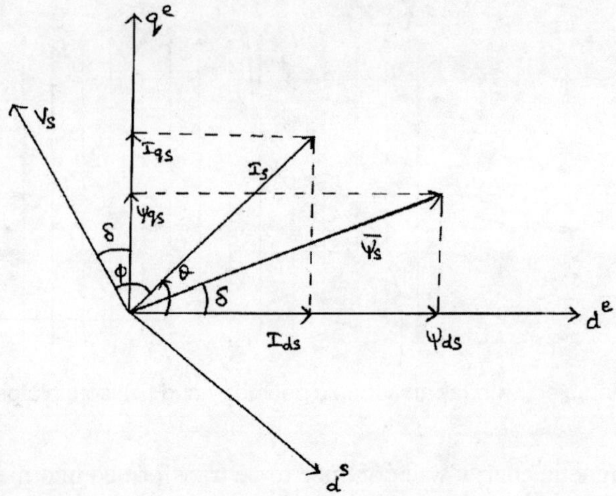

Fig.8.5 (b): Phasor diagram of the synchronous reluctance motor with d-axis as reference

8.1.3 (B) Torque-Speed and Torque-Load Angle Characteristics of Synchronous Reluctance Motor

In synchronous reluctance motors, the natural tendency of a ferromagnetic material to align itself with a magnetic field generates the reluctance torque. The synchronous reluctance motor is not self-starting on a fixed frequency a.c. supply unless the rotor is supplied with a squirrel-cage winding to allow starting by induction motor action. The rotor speed oscillates above and below its average value as the rotor speed approaches synchronous speed because the reluctance torque is superimposed on the induction motor torque. The instantaneous rotor speed rises to synchronous speed and the rotor locks into synchronism with the stator field if the load torque and inertia are not too high.

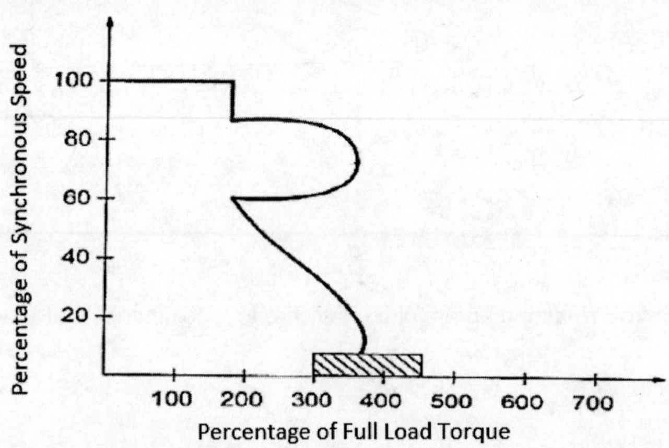

Fig.8.6: Torque-speed characteristics of synchronous reluctance motor

The torque-speed characteristics of a synchronous reluctance motor are shown in Fig.8.6. The motor starts as a two phase induction motor at between 300 and 400 percent of its full load torque (depending on the orientation of the rotor's salient pole axis with respect to the axis of the rotating magnetic field). By virtue of its saliency, the motor now operates as a synchronous motor once it reaches its maximum speed as an induction motor due to the reluctance torque pulling the rotor into step with the revolving field. The rotor can be pulled into synchronism with the pulsating single phase field as it gets closer to synchronous speed due to the reluctance torque. It is clear from Fig.8.6 that even when the torque is increased, the motor speed stays the same. However, the motor loses synchronism when the torque exceeds its maximum value. Up to a little bit more than 200% of its full load torque, the motor runs at a steady speed.

According to Fig.8.6, the motor speed stays constant even when the torque is increased. However, as the torque rises above its maximum value, the motor loses synchronism. Up to just over 200% of its maximum torque at full load, the motor runs at constant speed. A centrifugal switch opens with the beginning winding at 3/4th of the synchronous speed, and the motor continues to produce a single phase torque product using only its running winding. The reluctance torque is enough to bring the rotor into synchronism with the pulsating single-phase field as it approaches synchronous speed. Up to a little bit more than 200% of its full load torque, the motor runs at a steady speed. It will continue to run as a single phase induction motor up to 800% of its rated output if it is loaded beyond the draw out torque value.

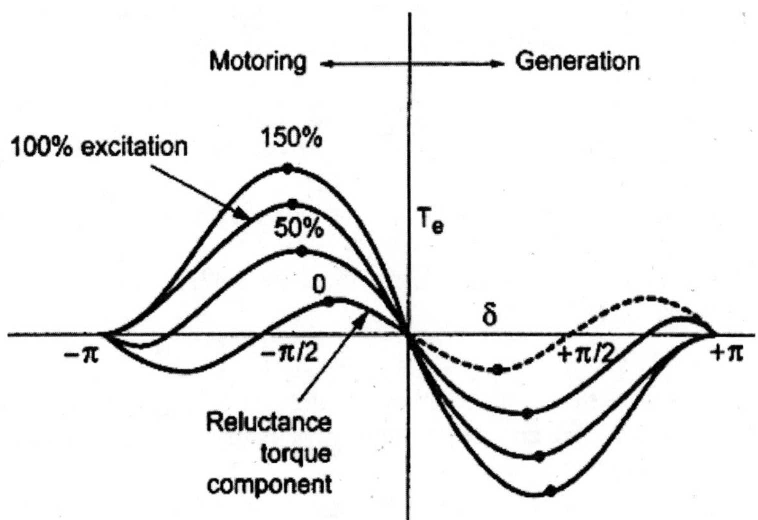

Fig.8.7: Torque- angle characteristics of salient pole machine

Plotting the previously mentioned equation for various field excitations results in the various torques (T_e) - angle curves for both driving and generating modes, as shown in Fig.8.7.

The dots in Fig.8.1.7 denote the steady-state limit, which is represented by the maximum points. It is clear from the equation that given a fixed excitation and torque angle, the produced torque remains constant if V_s/ω_e is kept constant (i.e., the supply voltage is altered proportionally to the frequency). However, the synchronous reluctance motor is defined as a motor with the same basic design as a salient pole synchronous motor but without a field winding on the rotor. As a result, the motor is not excited. As a result, the reluctance torque component is the lowest curve in the torque angle characteristic of Fig.8.7 generated for the salient pole machine, which corresponds to zero excitation or zero percent excitation.

The ideal synchronous reluctance machine has a rotor whose structure allows the inductance of the stator windings in the d-q reference frame to vary sinusoidally from a maximum value L_d [Direct inductance] to a minimum value L_q [Quadrature inductance] as a function of angular displacement of the rotor. This results in the reluctance torque component having the shape depicted in Fig.8.7.

8.1.4 Features of the Synchronous Reluctance Motor

Advantages:

- It has a reduced torque ripple.
- This motor's rotors can be built from inexpensive and highly durable materials.
- Standard PWM AC inverters can be used to operate this motor.
- This motor can withstand temperatures that are very high.
- This motor's structure is both sturdy and straightforward.
- This motor can operate at high speeds.
- Since there is no need for field excitation in this motor at zero torque, the losses caused by electromagnetic spinning are eliminated.
- Since demagnetization is not a concern, these motors are more reliable than permanent magnet motors.
- Due to this motor's simplicity, a multi-motor drive can use it to synchronously run multiple motors using a single power source.

Disadvantages:

- The cost of these motors is higher than the cost of an induction motor.
- Sensor less control and rotor position sensors must be used to synchronise speed with an inverter's output frequency.
- This motor is heavier and has a lower power factor than an induction motor.
- A variable frequency drive is used to power it.

Applications:

The synchronous reluctance motor has the following uses.

- Due to its affordability, robust construction, natural simplicity, etc., it is suitable for low-power applications such as fibre spinning machines.

- Timing devices, phonographs, control devices, recording instruments, etc. are some examples of applications where consistent speed is essential.
- It is utilised in place of pumps in conveyors as proportioning devices.
- These motors are utilised in synthetic fibre manufacturing equipment as well as turntables, regulators, synchronised conveyors, metering pumps, and regulators. Used in the process of film material otherwise continuous sheet.
- Used in folding, wrapping machines, and auxiliary time machine.

The synchronous reluctance motor overview is what this section is all about. Due to its higher efficiency and lack of magnets, this motor offers assurance by offering a sustainable environmental option to lessen its overall environmental impact. Its lower operating expenses will enable a quick payback. This product benefits from the great efficiency and dependability of synchronous motors as well as induction motors.

8.2 INTRODUCTION TO SWITCHED RELUCTANCE MOTOR

An electromechanical energy conversion tool that transforms electrical energy into mechanical energy is a switched reluctance motor. In contrast to DC motors, where energy conversion occurs as a result of the Lorentz Force Law, switching reluctance motors convert energy as a result of the principles of variable reluctance. The definition of resistance is the opposition of flow to the magnetic field. The resistance that obstructs current flow is its electrical equivalent.

The flow of the magnetic field is known as flux in the magnetic field, whereas the passage of charges is known as current in the electrical field. Magnetic poles develop on the stator and rotor of the reluctance motor as a result of the magnetic field's property.

Fig.8.8: Switched Reluctance Motor

The motor rotates as a result of the interaction between these poles. Switching circuits utilizing power electronics devices are utilized to change the magnetic poles.

As a result, the term "switched" has been employed. The structure of the Switched Reluctance Motor is depicted in Fig.8.8. The variable reluctance principle is responsible for the electromagnetic torque generated by reluctance motors. The poles must be generated with opposing charges in order to produce a varied reluctance; for example, if one pole of the stator is south one moment, it should be north the next. So that the rotor's rotation will be caused by a varied torque produced by this variable field. Power electronic components are utilized in a switching circuit to provide a changeable field. A spinning magnetic field is generated on the stator poles as a result of the switching circuits being activated.

8.2.1 CONSTRUCTION OF SWITCHED RELUCTANCE MOTOR

The machinery component and the switching circuit are the two main components of a switched reluctance motor. Stator poles and rotor poles are the components of a machine, such as a DC motor, in the machinery section. Silicon steel makes up the stator poles. Outside, the poles are projected. The number of poles is determined by the machine speed. We typically use 6 poles for speeds about 1000 rpm. Typically, ferromagnetic materials are used to wind the rotor. Additionally, these materials experience fewer eddy current losses and hysteresis. The stator construction has the rotor stamped into it. Hall-effect sensors are utilized to dynamically determine the rotor's position. This makes it possible to regulate the motor's speed. The motor's design is simple, robust, and requires minimal maintenance. It offers all-around excellent operation, and motor speed control is also made simple. Reluctance motors are one of the most often utilized motors as a result of these factors.

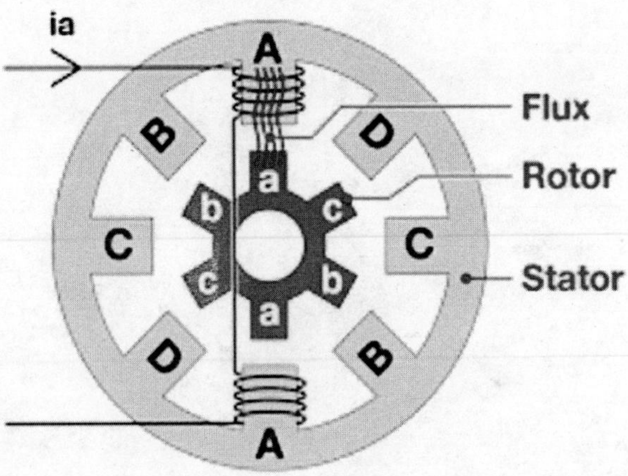

Fig.8.9: Outer part of stator poles and the inner parts of rotor poles

In Figure 8.9, the rotor poles are located inside while the stator poles are located outside. A single-phase supply excites the stator poles. Squirrel cage rotors are used to provide a rotating magnetic field for synchronous reluctance motors. A rotating magnetic field cannot be produced by a single-phase supply.

With a switched reluctance motor, switching circuits based on power electronics are used to create the rotating field. The reluctance principle states that magnetic flux always seeks to move along the path with the lowest reluctance, just as current seeks to move through the path with the lowest resistance. As a result, the reluctance between the poles of the stator and the rotor varies, and the rotor attempts to align itself with the axis with the least reluctance. The rotor is rotated in this process by a variable reluctance torque.

8.2.2 Switched Reluctance Motor Design

The design of a three-phase switching reluctance motor is depicted in the following Fig.8.10. It is supplied by a DC source, which a three-phase inverter converts into AC. Two MOSFETs or IGBTs serve as the switching components of each leg of the inverter. One phase of the inverters is connected to each phase of the stator. It is a three-phase inverter that changes DC into AC in response to the switches being triggered.

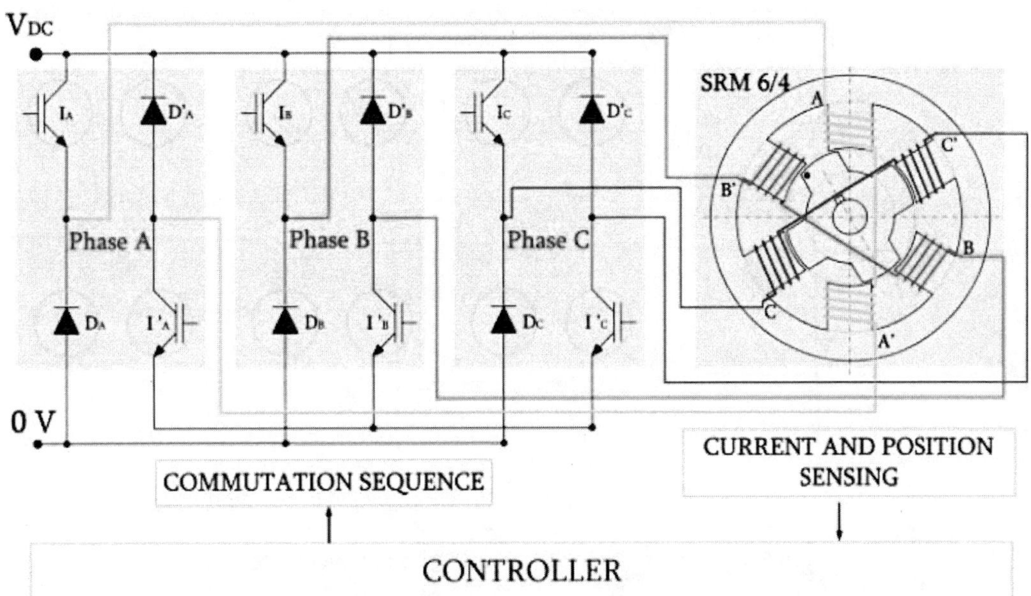

Fig.8.10: Switched reluctance motor design

According to the switching order, the switches are numbered. The switches A, B, and C make up the circuit's upper half. The circuit's lower half is made up of the switches A', B', and C'. Each switch from the upper to the lower half at a time is activated, and these two switches together make up one phase. These two switches are linked to the stator's two poles. The remaining phase is formed similarly by B, B' and C, C', which are connected to the stator's other poles.

8.2.3 Torque - Speed Characteristics of Switched Reluctance Motor

The mechanical subsystem design characteristics typically place limits on the maximum torque that can be created in a motor and the maximum power that can be transferred. By altering the duty cycle of the chopper, the torque can be changed for a given conduction angle. However, due to mechanical considerations, the greatest torque created is constrained to a specific value. The constant maximum torque area of operation is depicted in Fig.8.11, which also shows the torque speed characteristic of a switching reluctance motor (AB).

The torque-speed capability curve might deviate from the clock torque characteristics at very low speeds. It is challenging to limit the current without the aid of the motor's self-emf, and the current reference may need to be decreased, if the chopping frequency or the bandwidth of the current regulator are constrained. If extremely low windage and core loss allow for an increase in chopper losses, a larger torque can be obtained with a higher current. Naturally, in any part of the speed range up to ωb very considerably higher torque can be obtained under intermittent conditions.

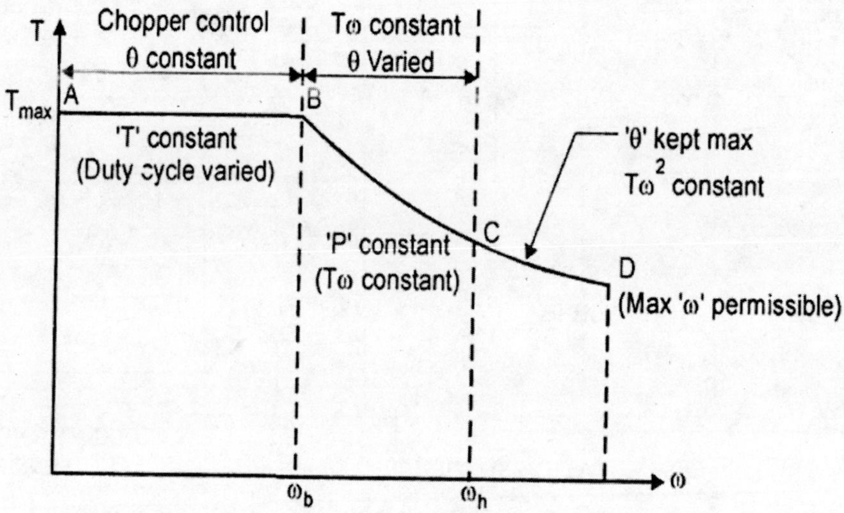

Fig.8.11: Torque - speed characteristics of switched reluctance motor

The torque below base speed is restricted by the motor current. The highest speed at which maximum current may be provided at rated voltage with fixed firing angles is known as the corner point, or base speed, or ωb. The maximal torque at rated voltage falls with speed squared if these angles are still held constant. The maximum current can still be driven into the motor over a sizable speed range, however, if the conduction angle is increased (i.e.,) θ on decreased. This keeps the torque at a greater level to keep the power characteristic consistent. But the core losses and windage losses increases with the speed.

Thus the curve BC represents the maximum permissible torque at each speed without exceeding the maximum permissible power transferred. This region is obtained by varying θ_D to its maximum value θ_D max. θ_D is dwell angle of the main switching devices in each phase. Point C corresponds to maximum permissible power; maximum permissible conduction angle. The highest possible torque that can be applied at each speed without exceeding the maximum permitted power transfer is represented by curve BC. By adjusting θ_D until it reaches its maximum value, θ_D max, this region is obtained, where θ_D is the dwell angle of each phase's primary switching components. Maximum power and maximum conduction angle are corresponding to Point C.

8.2.4 Characteristics switched reluctance motor

The characteristics of the switched reluctance motor can be summarised as

- The switched reluctance motor could be single-phase or three-phase.
- Managing speed is simple.
- High speed can be attained by modifying the triggering circuits.
- An inverter can be used to operate it with a DC source as well.
- One phase's control is separate from the control of the other two phases.
- The feedback diodes can be used to recover the unused energy that was given to the a motor. This increases efficiency.
- By adjusting the switching devices' firing angles, various speeds can be achieved.

Advantages

The following are some benefits of a switching reluctance motor.

- Because the unused energy can be returned to the source via the feedback diodes, this motor is more efficient than a DC motor.
- This motor's speed can be easily controlled because the supply phases can be adjusted to a great amount, allowing for a smooth change in speed.
- The motor can be energized both by an AC and a DC supply.
- Stator poles require minimal maintenance because they are durable.
- The switches' voltage rating is likewise lower, which makes the machine more inexpensive.

Disadvantages

The disadvantages of the motor are as follows

- The machine's existence of ripple torque is its primary drawback. Unwanted interferences, also known as harmonics, between the rotor and stator are what produce the ripple torque. Unwanted losses and torque ripple are results of the harmonics. But it can be avoided by utilizing cutting-edge modulation methods like pulse width modulation or space vector modulation.
- Because switches are present, they frequently experience wear and tear issues. Therefore, we must frequently replace them. But because they are less expensive, the system's efficiency won't be affected.

- Switching losses also reduce productivity. The switches typically have moderate conduction losses and high switching losses.

Applications
Due to its versatility, switched reluctance motors are often used in

- Robotics
- Aviation
- Industrial Applications
- Automation
- Washing machines
- Vacuum cleaners and fans

This section discussed switching reluctance motors in action and learned how they work. These motors have become more well-liked as a result of the development of power electronic devices. They have some restrictions nevertheless because they are dependent on switching devices. The operation is additionally enhanced by the usage of hall-effect based sensors. It would be intriguing to consider how to regulate the speed of a closed-loop switching reluctance motor. Which controller would perform more effectively?

8.3 BRUSHLESS DC (BLDC) MOTOR

Despite the fact that brushed motors existed prior to the development of brushless motors, German physicist Ernst Werner Von Siemens created the first brushed DC motor in 1886. Because brushed motor use is inefficient and expensive to maintain, there are several difficulties. In 1962, the invention of brushless type took place by T.G. Wilson and P.H. Trickey, while these difficulties were being considered. The design of this brushless motor paved the way for increased productivity and time savings in several industries.

To address some challenges, a BLDC motor, also referred to as a brushless DC motor, is a motor that operates without brushes. A brushed one and a BLDC operate differently depending on whether brushes are present or not. Brushes are used to transfer or gather current to the commutator in a typical machine. However, sparks result when the brushes are used. It slides between the brushes as and when the rotor revolves. During the rotation of the shaft, the brushes ignite. Life expectancy also decreases as a result of this reduction in efficiency. In order to overcome these drawbacks, a BLDC was created that operates much more smoothly than a typical one. Sparking is also avoided because the brushes are not used. This lengthens the motor's lifespan and improves its performance.

Microcontrollers are capable of controlling all of the common motor types, each of which has a unique use. Today's motors are all switching to brushless DC technology. Although DC motors have earned their popularity over the years, the BLDC option is significantly more advantageous due to advancements in microcontroller affordability and motor design. Every common type of motor has a particular application and can be controlled by a microcontroller. We can better comprehend the operation of the BLDC motor by using the DC motor as a reference, as most people have a good intuitive understanding of the way a DC motor works.

8.3.1 Brushed DC Motors Review

Within a stationary magnetic field, a DC motor (brushed) with winding assembly (armature), brushes, and commutators switches current to separate windings in the proper ratio to the external permanent magnet field is shown in Fig.8.3.1. The pros and cons of this motor are

Pros:

- The easiest electronic control eliminates the requirement for controller commutation.
- Needs just four power transistors.

Cons:

- Speed control necessitates the use of a sensor.
- The brushes and commutator generate sparks and eventually wear out.
- Sparks reduce max power.
- It is difficult to remove heat from an armature.
- Poor power density

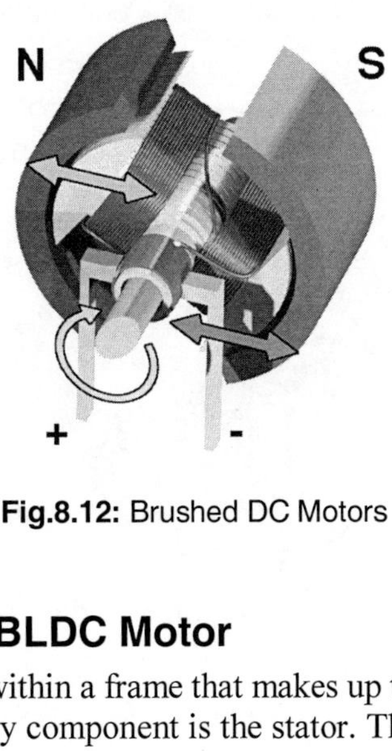

Fig.8.12: Brushed DC Motors

8.3.2 Construction of BLDC Motor

The rotor and stator are fixed within a frame that makes up the device. The revolving component is the rotor, while the stationary component is the stator. The frame serves as both a support for the internal peripherals and a shield against the outside elements. Permanent magnets are used to hold the rotor in place, and the stator's coils serve as an electromagnet. The number of poles is determined by operational needs. The slots in the stator are used to house the coils. The number of poles determines how many coils should be used.

These coils are energised in accordance with the requirements. A brushless direct current motor is made up of a permanent magnet rotor and stationary windings as shown in Fig.8.3.2 . The Pros and Cons of Brushless DC motor are

Pros:

- There are no brushes or commutators to wear out.
- There are no sparks or extra friction.
- More efficient than a direct current motor
- Higher speed than DC motor
- Better power density than a DC motor

Cons:

- Commutation requires a rotor sensor or sensorless approaches.
- 6 power transistors are required.

Brushless DC motors are DC motors that are built from the inside out and do not have brushes. It has the benefit of being a permanent magnet, which means that all of the power applied may be utilised for torque. It contains windings on the outside, allowing heat to travel through the outer wall and be readily dissipated. This is something that a DC motor cannot achieve. In addition, the absence of brushes eliminates the friction and sparks that hinder DC motors from operating at high speeds. However, the requirement for six transistors instead of four implies that the electronics cost will be higher, and the electronic commutation will make it more complicated.

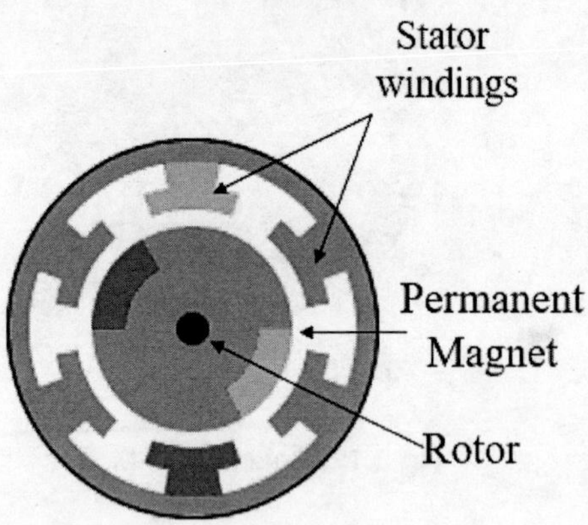

Fig.8.13: Construction of Brushless DC motor

8.3.3 Brushed DC Commutation

The brushes and armature switch the windings in the armature to DC power. Every winding experiences a positive voltage, followed by a disconnect, and finally a negative voltage. The armature's field interacts with the static magnet, causing torque and revolution.

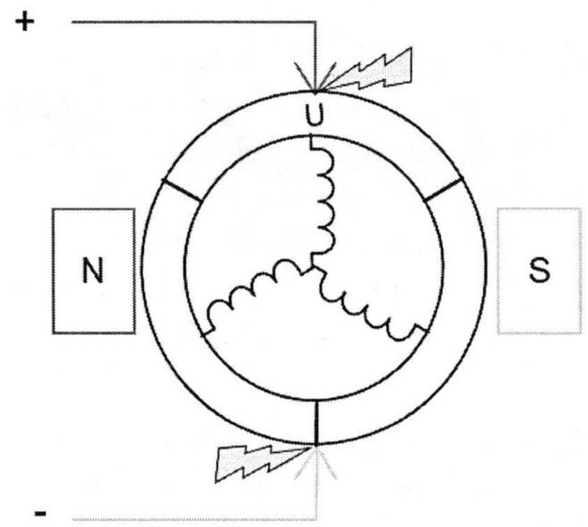

Fig.8.14: Inside-out model of the BLDC motor.

Despite the fact that actual DC motors can differ greatly, this one is a good model. Regardless of whether this variant with three windings in the armature exists, DC motors are substantially more probable to have more commutator bars, windings, and brushes, in particular when they get larger. The idea of the windings switching DC voltage remains unchanged, but the waveform becomes more intricate. This model is shown in Fig.8.14 as a straight forward illustration that best illustrates the BLDC motor's "inside-out" design Take notice of the sparks at the brushes; this is where ionisation may result in a "ring fire—a brush-to-brush short circuit.

8.3.4 Three-Phase Bridge to Drive BLDC Motor

A DC motor built from the inside out but without brushes or commutators is what is known as a brushless DC motor. Transistors are used in place of the mechanical switches, and the windings are shifted from the armature to the stator. From the outside, the magnet is moved inside to become the rotor.

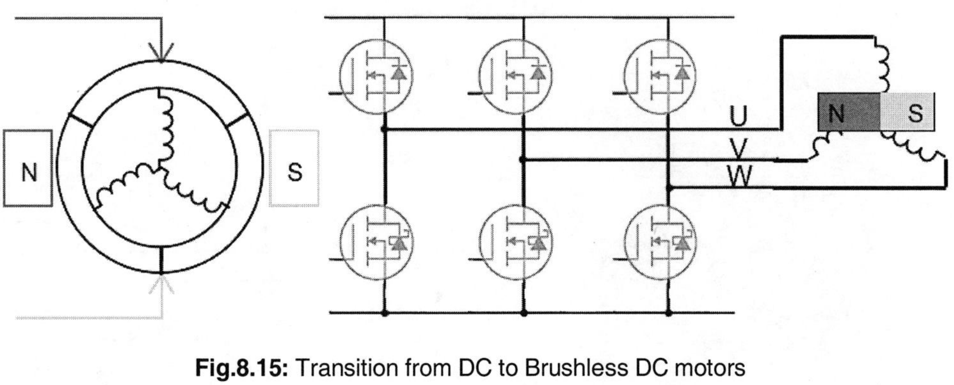

Fig.8.15: Transition from DC to Brushless DC motors

Fig.8.15 depicts the transition from DC to Brushless DC motors. The six-transistor (3-phase) bridge's replacement of the brushes and commutators in particular as shown in Fig.8.16. In practice, iron laminations carry the magnetic fields, closing the magnetic loops and minimising the air gaps. All motors must have this to operate at their peak efficiency. Here, the transistors are MOSFETs, which are a well-liked option at lower voltages. Insulated gate bipolar transistors are employed in higher-voltage applications. To manage inductive kick or flyback currents, each transistor must have a diode connected in parallel with it in either scenario.

Hall sensors sense magnetic fields and are able to be employed to determine the angle of a rotor. For each sensor, the output is a digital 1 or 0, depending on the magnetic field nearby. As indicated in 8.17 each Hall sensor is positioned 120 degrees apart on the back of the motor. The Hall sensors output logic bits that show the angle as the rotor rotates. At any given time, one winding is connected to positive via an upper transistor, one is connected to negative via a lower transistor, and one is shut off.

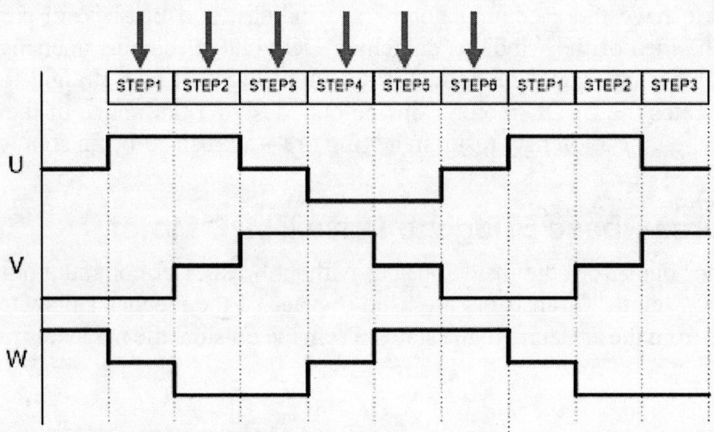

Fig.8.16: Six-step Commutation

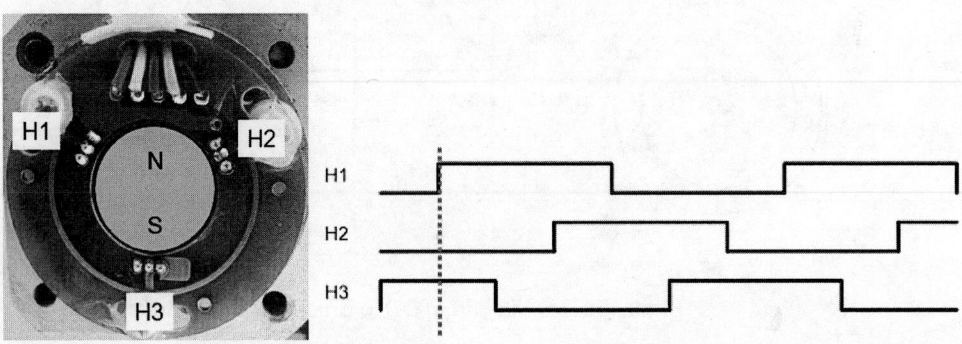

Fig.8.17: Hall Sensors and its output

This is also known as trapezoidal or 120- degree commutation. Though the 120-degree phrase is not evident to most people, it simply indicates that one winding, or phase, is shut off at any given time, resulting in just two-thirds of the current flowing. All windings are used in 180-degree commutation. When all three sensors are combined as shown in Fig.8.17, they create six distinct logic combinations or steps. The motor phase combinations are deciphered from these three bits. The Hall sensor, named for the scientist who discovered the phenomenon, produces very low-level voltage changes in response to a nearby magnetic field. The majority of sensors on the market have the amplification and comparator required to convert them to digital devices. They increase the cost and length of the motor, and they are exposed to the hostile environment of the motor, limiting its heat capability and power. The remaining structure of the motor, which is constructed of copper, iron, and magnets, can generally withstand far higher temperatures than silicon.

Based on their binary states, the three hall bits are assigned a number. The first step is binary 101, the second is 001, and so on. Filling out a table of six choices, plus the two more invalid states of 000 and 111, yields all of the possible combinations. Then, when the Halls are equal to 001 (decimal 1) or 101 (decimal 8), turn on the U-phase upper transistor. This order is only acceptable when the motor is only turning one way. For reverse operation, create a new table in which the opposite transistor (swap upper and lower) is turned on for each state. These devices are shown as totally on or off for clarity; however, they are actually PWM at a certain duty cycle almost every time. The hall sensor and the inverter output is shown in Fig.8.18.

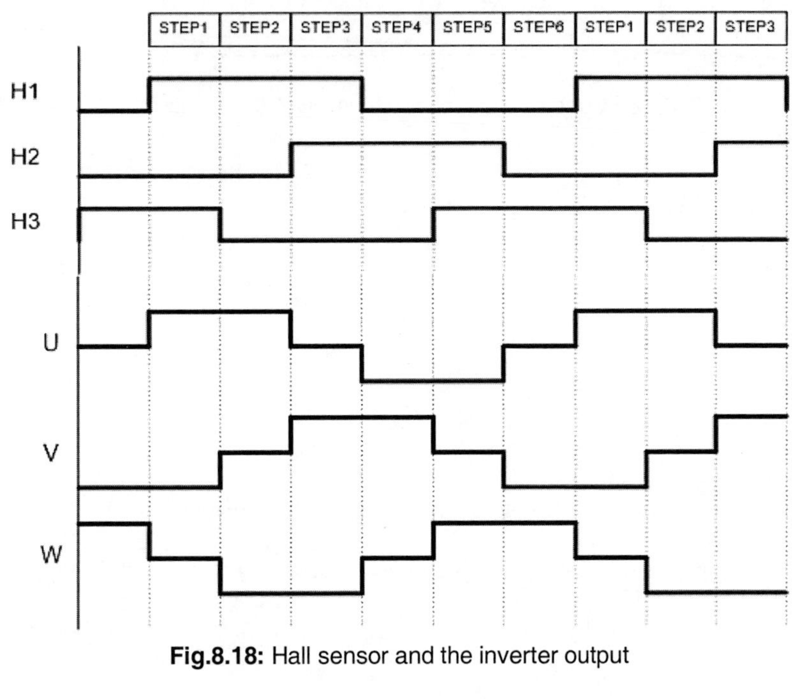

Fig.8.18: Hall sensor and the inverter output

8.3.5 Torque-Speed Characteristics of PMBLDC Motor

Assume that commutation is accurate, the phase current waveform is ideal, and the converter is powered by an ideal voltage source, V. Then, V = E + R I

When the phase resistance is low, the properties are similar to those of a DC shunt motor. The speed can be changed by altering the voltage (V). Chopping, or PWM, is used to control voltage. There are limits to both continuous and discontinuous operations. Fig.8.19 (a)shows the Speed-Torque characteristics of BLDC motors. Heat transmission and temperature rise occur during the continuous limit, while the maximum rating of the semiconductor switch happens during the discontinuous limit. Figure 8.19 (b) depicts the T-N characteristic family for various constant supply voltages, Current, torque, supply voltage, and speed should all be within acceptable limits.

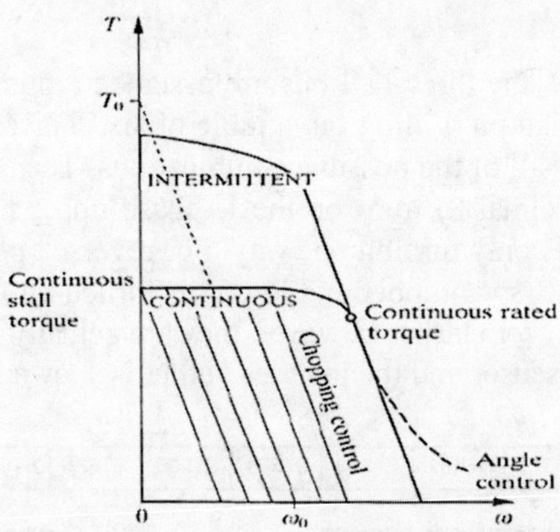

Fig.8.19 (a): Speed-Torque characteristics of BLDC motors

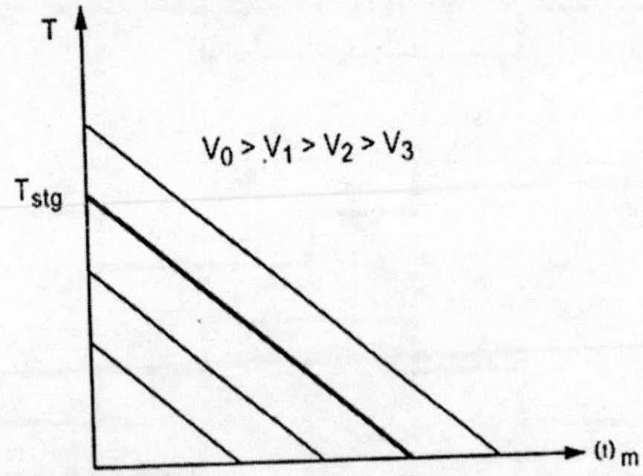

Fig.8.19 (b): Family of T-N characteristics for various constant supply voltages.

Line AB parallel to x-axis represent the maximum torque developed. Line FG represent the T-N characteristics for maximum permissible voltage and line DH perpendicular to x-axis represents maximum permissible speed. The area OABCDO shown in Fig.8.19 (c)is the permissible region of operation.

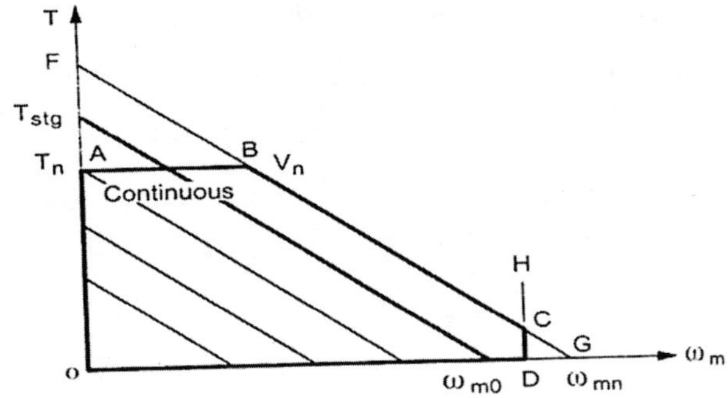

Fig.8.19 (c): Permissible region of operation.

Types of BLDC Motor

BLDCs are classified into two categories: They are classified as in-type and out-type. It is determined by the structure; for example, if the rotor is designed to rotate outside, it is out-type. The rotor is designated IN-type if it is designed to rotate internally. Out-type is the most commonly used type.

Controlling Speed of BLDC Motor

The speed of this type can be controlled relatively easily because it has a controller. The voltage may be adjusted to control the speed. The speed of the coils can be changed depending on the amount of voltage applied to energise them. To lessen the speed of the motor, the voltage provided to the coils is reduced.

Advantages

- Highly reliable
- Smoother operation
- High power output
- Can be used for higher speed applications
- No brush is required
- No sparking
- Light in weight
- Less noise

Disadvantages

- Complex operation
- Requires maintenance
- A controller is required

- Costly
- Sensors are required

Applications

The applications of PMBLDC motors are given below.

- Industrial drives (fans, pumps, blowers, handling systems)
- Machine tools
- Servo drives
- Automation process
- Internal transportation systems
- Robots
- Air conditioning systems
- Washing machines
- Vacuum cleaners
- Security systems
- Elevators
- Light railways
- Street cars
- Electric road vehicles
- Air craft flight control

A BLDC motor was defined in this section. It is a motor that does not need brushes to offer a smoother as well as more effective operation. Its construction, operation, the difference between a normal and brushed type, speed control, uses, flaws, and application were all discussed. What is a commutation in DC machines, and why is a commutator employed, is a question for the readers.

8.4 PERMANENT MAGNET SYNCHRONOUS MOTORS

Nearly two decades before the induction motor was discovered, synchronous motors were first developed in the year 1869. The induction motor was created in the first decades of the 20th century, and its uses were greatly expanded. While the synchronous motor's range did not expand until the early 1920s. It took almost 8 years to advance the required initial capability through interaction with the induction motor, include synchronous motor uses for power factor modification, and ultimately eliminate the manual initiation restriction by developing direct control for synchronous motors. Permanent Magnet Synchronous Motor (PMSM) is a type of synchronous motor used as a permanent magnet for the creation of the excitation field. Because it produces a sine-shaped flux distribution in the space between the rotor and stator components, this sort of permanent sine wave motor is also known as a brushless type. Even the current wave produced by that device has a sine wave pattern.

PMSMs offer higher torque for small frame sizes and no rotor current since they are made of permanent magnets. These devices also offer a good range of power-to-size ratios, enabling straightforward designs with little torque loss. There must be commutated similarly to BLDC motors, but due to the design of the windings, the signals must be in fine shape to provide better performance. The PMSM devices are mostly used in controllers like digital signal processors or microcontrollers since they are built with sophisticated controlling algorithms. Fig.8.20 depicts the permanent magnet synchronous motor's construction.

The machine is primarily made up of a revolving element called a rotor and a stator that is stationary. Axial air gap devices' laminations are most likely produced by rolling them with soft steel strips. Different lamination portions act as teeth gaps where armature windings are also present, and the yoke section ends up becoming the magnetic route. The rotor of a motor is typically found inside the stator; however occasionally externally positioned rotors may exist. The permanent magnets that make up the rotor sections have strong coercive forces. Synchronous motors are divided into electric motors with and without prominent pole rotors based on the rotor's design. Compared to an electric motor with a salient pole rotor, which has inductances that are not equal, L_d L_q, the motor with a non-salient pole rotor has equal quadrature and direct inductances, or $L_d = L_q$. Fig.8.21 displays several motor configurations dependent on the design of the rotor.

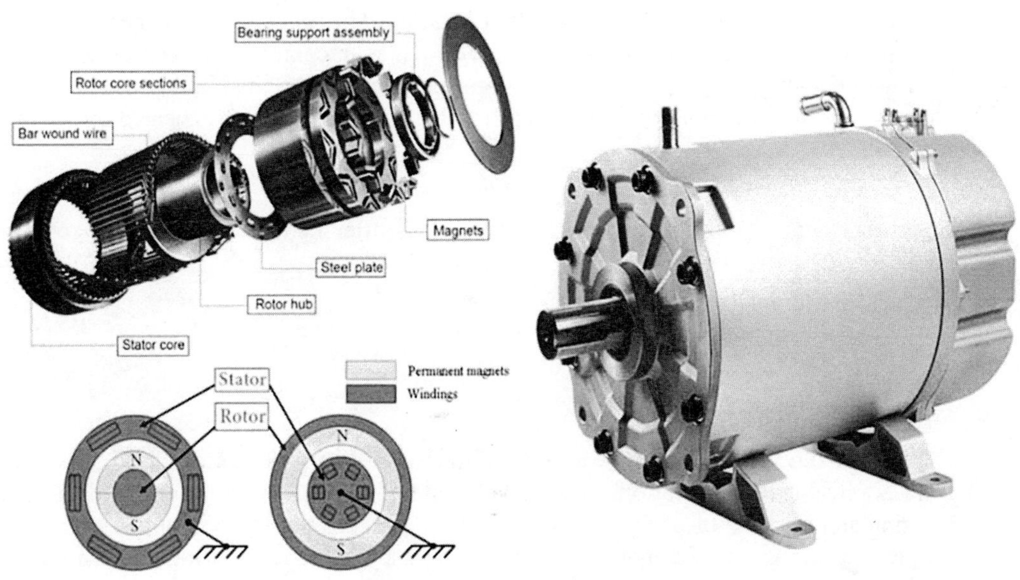

Fig.8.20: Construction of PMSM

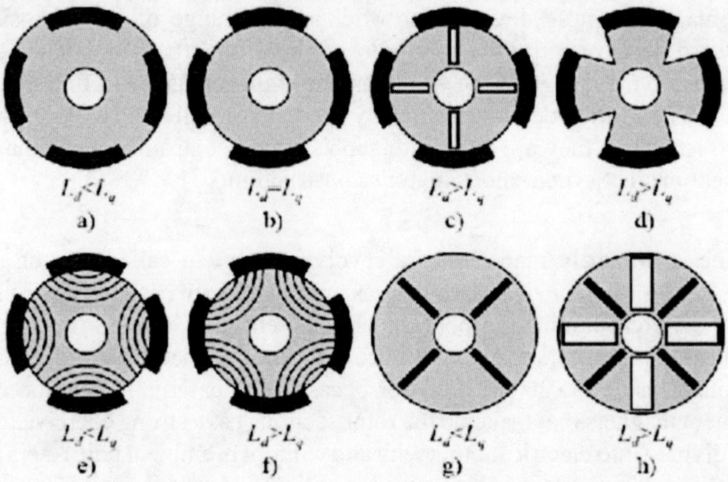

Fig.8.21: Motor Configurations based on rotor design

Additionally, there are two types of PMSM dependent on the architecture of the rotor: surface PMSM and inner PMSM. These contain a core part with windings and a stator section with an exterior frame. The design of this kind of rotor is frequently seen in 2-phase and 3-phase winding schemes. The second form of PMSM is one with scattered winding and one with concentrated winding, and it is dependent on the design of stators. The number of slots per pole and phase in the distributed winding is in the range of Q = 2, 3.... k. In contrast, the focused winding contains slots with equal spacing over the whole stator perimeter since the number of slots per pole and phase is Q = 1. The two coils that make up the winding can be connected either in series or parallel.

Additionally, the PMSM is divided into two kinds after being attached to the rotor core by a permanent magnet:

- Surface Mounted PMSM
- Interior/Buried PMSM

Surface Mounted PMSM

Here, the magnet is placed on the surface of the rotor. As the design of this kind is not robust, it is not applicable for applications those function at high speed. As the magnet and air permeability are similar, this kind of design creates a regular air gap. So, there is no presence of reluctant torque. Regarding this, the motor's dynamic efficiency is higher, making it appropriate for high-performance machine tool drivers and the robotics industry.

Interior/Buried PMSM

In this case, the permanent magnets are positioned inside the rotor rather than on its surface. This enhances device strength and so used for applications that operate at high speed. As because of the salience feature, there exists reluctant torque in the PMSM.

8.4.1 Working of Permanent Magnet Synchronous Motor

The permanent magnet synchronous motor working is dependent on how the rotor's static magnetic field and the stator's revolving magnetic field interact. The principle of revolving magnetic field in the stator section of the motor is similar to the 3-phase induction motor. As per Ampere's principle, the rotor's magnetic field operating with the synchronous AC of the stator windings generates a torque that allows the rotor to start revolving. The permanent magnets which are located on the rotor of the motor generate a static magnetic field. When the speed of the stator field and rotor's rotational speed synchronizes, then the rotor poles link up with the revolving magnetic field of the stator.

As per this, the PMSM device will not start automatically. It has the self-starting ability when the device has a direct connection with the 3-phase current network which means when the current frequency is in the range of 80 Hz.

Torque-speed characteristics

The torque-speed characteristics of BLPM sine wave motor is shown in Fig.8.4.3.For given V_C, I_C (maximum permissible voltage and current) torque remains constant from low frequency to corner frequency f_0. Any further increase in frequency decreases the maximum torque. The shaded part represents the permissible region of operation in torque speed characteristic.

If the speed is raised past point D in the torque speed characteristics, there is a risk of over current since the back emf E_b keeps growing while the terminal voltage stays constant. After that, the current travelling from the motor back to the source is virtually entirely reactive. Due to losses in the motor and converter, the q axis current and torque are both minimal. Thus, the power flow is turned around. This mode of operation is only feasible if the motor exceeds the converter's capacity or is powered by an external load or prime mover.

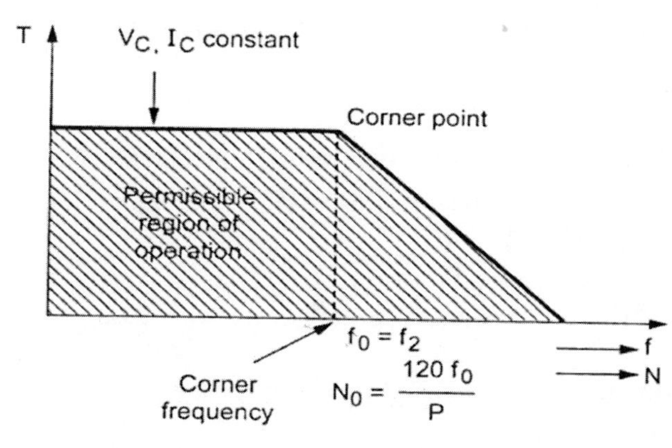

Fig.8.22: Torque-speed characteristics

Comparison between PMSW and BLDC

Table 8.1: Comparison between PMSW and BLDC

Permanent Magnet Synchronous Motor	Brushless DC Motor
It is a synchronous machine	It is a synchronous machine
The input is sinusoidal currents	The device is fed with direct currents
This device is suitable for higher speed levels	This device is suitable for lower speed levels
It generates less noise	It generates high noise when compared with PMSM
Complicated controlling algorithms	The device can be constructed with simple controlling algorithms
It has the possibility of 3 phases ON at a single instance	For a BLDC, only 2 phases can be in ON state at the same time
Minimal core losses	It has high core losses as because of harmonic content

Benefits and Drawbacks of PMSM

The advantages of permanent magnet synchronous motor are:

- It is capable of eradicating field copper losses.
- Enhanced power density
- Minimal rotor inertia
- Robust motor architecture
- Smooth torque and high dynamic performance are the outputs.
- Delivers increased efficiency

Permanent magnet synchronous motor drawbacks include:

- The PMSM device is more expensive than other motor types available today.
- It affects demagnetization.
- These motors do not self-start.
- Loss of field flux control's adaptability

Permanent Magnet Synchronous Motor Applications

The applications of PMSM are

- Numerous home appliances, including air conditioners, fans, lifts, and compressors
- Applicable to factory automation

- Vacuum cleaners
- Applied to robotic systems
- Shutters and gates
- Surveillance platforms

8.5 STEPPER MOTOR

A stepper motor is a type of spinning electric motor having discrete angular motions of the rotor caused by control signal pulses. The stepper motor served as the servo motor's successor. Stepper (pulse) motors convert the control signal in the form of a pulse sequence directly (without the use of a feedback sensor) into a fixed angle of shaft rotation or linear movement of the mechanism proportional to the number of pulses. This circumstance simplifies the drive system and replaces the closed-loop servo drive (servo) system with an open-loop one, which has benefits such as lower device cost (fewer elements) and increased accuracy due to stepper motor rotor fixation in the absence of signal pulses. As a result, servo motors are employed in tasks requiring high performance (accuracy and speed). In other circumstances, stepper motors are commonly employed due to their inexpensive cost, ease of control, and high accuracy.

8.5.1 The Construction of Stepper Motor

A stepper motor, similar to any other spinning electric motor, is made up of two parts: a rotor and a stator shown in Fig.8.23. The stator is the static component, whereas the rotor is the revolving component. Because the rotor lacks slip rings and a commutator, stepper motors are both dependable and affordable. The rotor is equipped with either salient poles or tiny teeth. The rotor of the reluctance stepper motor is made of soft magnetic material with salient poles. A permanent magnet rotor is used in a permanent magnet stepper motor. The hybrid stepper motor comprises a composite rotor with soft magnetic material teeth (pole tips) and permanent magnets.

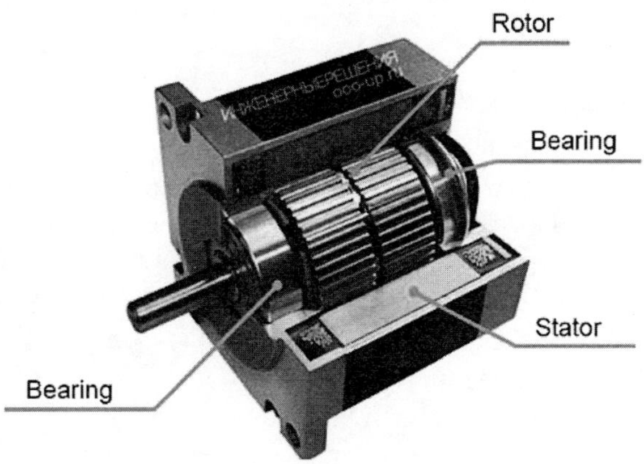

Fig.8.23: Hybrid stepper motor

Rotating a de-energized motor can reveal if the rotor has permanent magnets or not; if there is a holding torque and/or pulsation during rotation, the rotor contains permanent magnets. A stepper motor's stator includes a core with salient (protruding) poles that are often composed of laminated stamped electrical steel sheets to prevent eddy currents and heat build-up. A stepper motor's stator typically contains 2 to 8 phases. Power is not displayed in the stepper motor's specifications because it is not designed for continuous spinning. In comparison to other electric motors, a stepper motor has a modest power output. The rotor pitch (step), or the angle of rotor rotation corresponding to one pulse, is one of the defining parameters of a stepper motor. When the control pulses change, the stepper motor takes one step per unit of time. The step size is determined by the motor's architecture, which includes the number of windings, poles, and teeth. The step size might range from 90 to 0.78 degrees depending on the motor's design. Using the right control method, you can still achieve a step reduction in half with the help of the control system.

8.5.2 Characteristics of Stepper Motor

The important characteristics of stepper motors are

- The rotor revolves in distinct angular increments.
- The average motor speed is proportional to the rate at which the pulse command is issued.
- When the command pulse rate is low, the rotor rotates in steps.
- At a high command pulse rate, the rotor moves smoothly.
- Torque of the motor varies from 1 μN-m to 40 N-m.
- Power output range varies from 1 W to 2500 W
- High Torque-to-inertia ratio

8.5.3 Types of Stepper Motors

The stepper motors are classified as following types

a. Variable reluctance (VR) stepper motor
b. Permanent magnet stepper motor
c. Hybrid stepper motor

a. Variable reluctance (VR) stepper motor

A synchronous reluctance motor is a variable reluctance stepper motor. A reluctance stepper motor's stator typically has 6 salient poles and 3 phases (a pair of poles per phase), while the rotor has 4 salient poles and a step of 30 degrees. A switched-off variable reluctance stepper motor, unlike conventional stepper motors, does not have a holding (braking) torque as the shaft rotates. The rotor and stator teeth form the magnetic circuit's reluctance, which varies with the rotor's angular position.

Types of variable reluctance motor

i. Single-stack type
ii. Multi-stack type

i. Single stack VR stepper motor

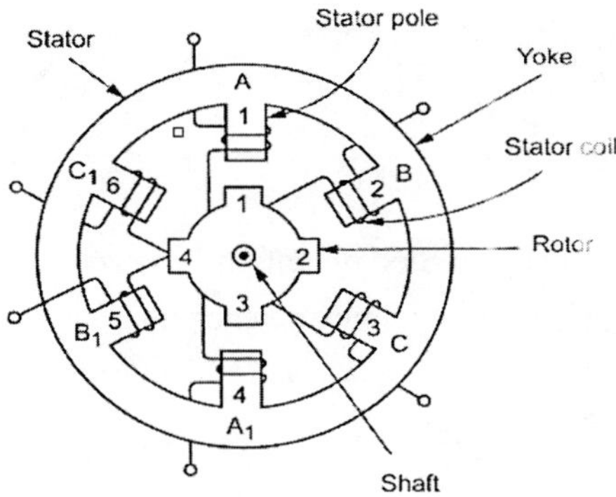

Fig.8.24: Construction of Single stack VR stepper motor

The construction and schematic diagram of single stack VR stepper motor is shown in Fig.8.24 and Fig.8.25. There are no permanent magnets in the stator. The poles are salient types and oncentrated windings are used. Field windings are carried by each pole. Even numbers of poles in the stator and in series phase windings are present, opposite poles are coupled. There are no permanent magnets in the rotor. The number of poles on the stator and rotor differs.

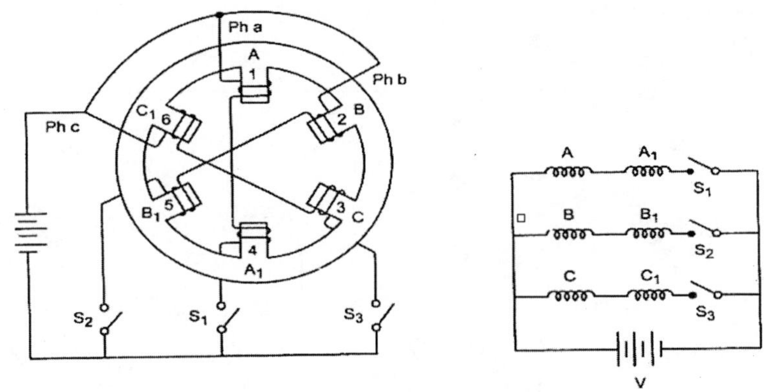

Fig.8.25: Schematic of VR stepper motor

Operation

Rotor strives for the lowest possible reluctance position.

$$\text{Step angle, } \beta = \frac{N_S - N_r}{N_S \cdot N_r} \times 360^0$$

$$\text{Step angle} = 360^0/mNr$$

Where

N_s-No. of stator poles or stator teeth

N_r-No. of rotor poles or rotor teeth

m-no. or stator phases

Motor has the following modes of operation.

Mode 1: one – phase ON or Full Step operation

Mode 2: Two – phase ON

Mode 3: Half – step operation

Mode 4: Micro stepping

Mode 1: one – phase ON or Full Step operation: Shown in Fig.8.26 (a) and (b)

Only one phase is energized at any time and each phase is excited by using switches.

Phase-A excited

A1 A2 gets aligned with rotor teeth 1 & 3.

Phase-B excited

B1 B2 gets in line with rotor teeth 4 and 2.

The rotor rotates at an angle of = 30.

Clockwise direction

Phase-C excited

C1 C2 aligned with rotor teeth 3 and 1.

The rotor rotates at an angle of $\theta = 30$.

Rotor total movement = 60°.

Clockwise direction.

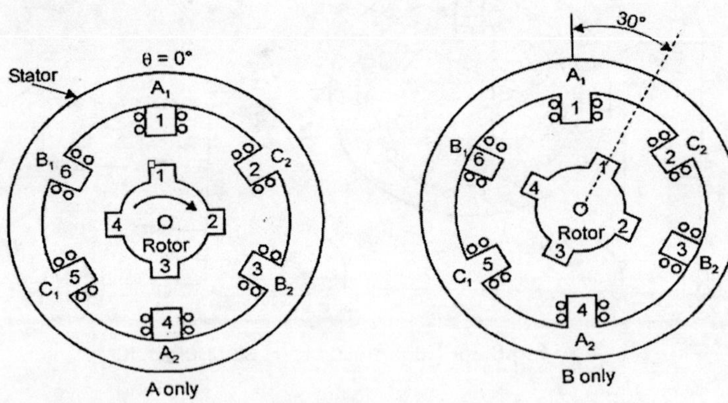

Fig.8.26 (a): One-phase ON mode (Phase-A & Phase-B)

Phase-A excited

- A1 and A2 are aligned with rotor teeth 4 and 2.
- The rotor rotates at an angle of $\theta = 30$.
- Rotor total movement = 90°.
- Clockwise direction

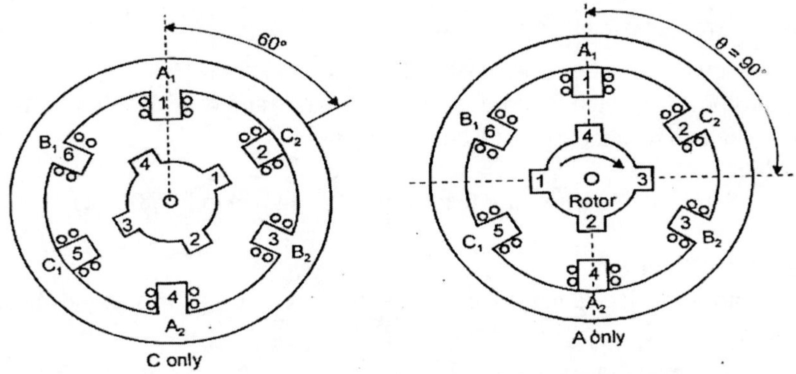

Fig.8.26 (b): One-phase ON mode (Phase-C & Phase-A)

Stator sequence: ABCA – clockwise direction

Stator sequence: ACBA – anti-clockwise direction

Table 8.2: Truth table for one-phase ON mode

Clock Wise Direction				**Anti-Clock Wise Direction**			
Ph-A	**Ph-B**	**Ph-C**	**Degree**	**Ph-A**	**Ph-B**	**Ph-C**	**Degree**
+	-	-	0°	+	-	-	0°
-	+	-	30°	-	-	+	30°
-	-	+	60°	-	+	-	60°
+	-	-	90°	+	-	-	90°
-	+	-	120°	-	-	+	120°
-	-	+	180°	-	+	-	180°
+	-	-	180°	+	-	-	180°
-	+	-	210°	-	-	+	210°
-	-	+	240°	-	+	-	240°
+	-	-	270°	+	-	-	270°
-	+	-	300°	-	-	+	300°
-	-	+	330°	-	+	-	330°
+	-	-	360°	+	-	-	360°

Mode 2: Two – phase ON operation

- At any given time, two phases of stator windings are energised.
- The rotor is subjected to torque from both phases and can come to rest at any point halfway between two neighbouring full step positions.

At initial,

A1 and A2 are aligned with rotor teeth 1 and 3 as shown in Fig.8.27(a).

Phase-A & B excited

- The rotor is subjected to torque from both the A and B phases as shown in Fig.8.27(b).
- Rotor rotates at an angle of $\theta = 15^{\circ}$.
- Clockwise direction

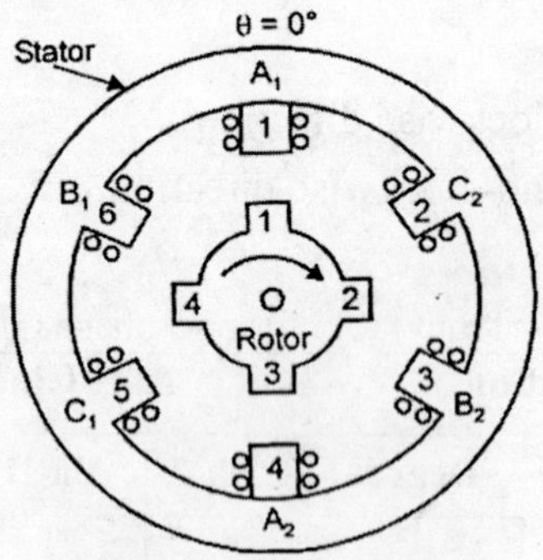

Fig.8.27(a): Initial position

Phase-B & C excited

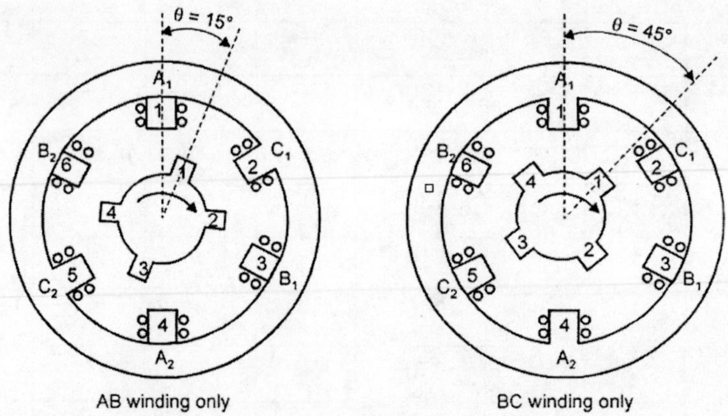

Fig.8.27 (b): Two-phase ON mode (Phase-AB & BC)

- The rotor is subjected to torque from both the B & C phases.
- Rotor rotates at an angle of 30°.
- Rotor total movement is $\theta = 45^{\circ}$.

- Clockwise direction

Phase-C & A excited

- The rotor is subjected to torque from both the C and A phases as shown in Fig.8.27(c).
- Rotor rotates at an angle of 30°.
- Rotor total movement is θ = 75°.
- Clockwise direction.

Phase-A & B excited again

- Rotor experiences torque from both A and B phases
- Rotor moves through an angle of 30°.
- Total movement of rotor is θ = 105°.
- Clockwise direction.
- Stator sequence : AB, BC, CA – clockwise direction
- Stator sequence : AC, CB, BA – anti-clockwise direction
- Torque developed in two-phase on mode is more than torque developed in one-phase on mode.
- Rotor transient damping occurs faster in this mode than in two-phase on mode.

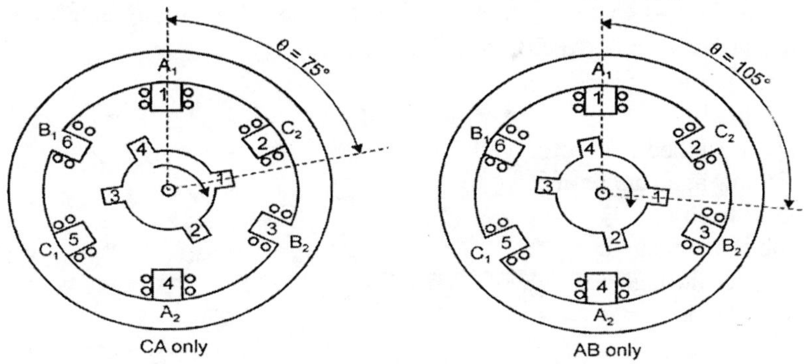

Fig.8.27 (c): Two-phase ON mode (Phase-CA & AB)

Table 8.3: Truth table for Two-phase ON mode

Clockwise Direction				Anti-Clockwise Direction			
Ph-A	Ph-B	Ph-C	Degree	Ph-A	Ph-B	Ph-C	Degree
+	+	-	15°	+	-	+	15°
-	+	+	45°	-	+	+	45°
+	-	+	75°	+	+	-	75°
+	+	-	105°	+	-	+	105°
-	+	+	135°	-	+	+	135°
+	-	+	165°	+	+	-	165°

Clockwise Direction				Anti-Clockwise Direction			
Ph-A	Ph-B	Ph-C	Degree	Ph-A	Ph-B	Ph-C	Degree
+	+	-	195°	+	-	+	195°
-	+	+	225°	-	+	+	225°
+	-	+	255°	+	+	-	255°
+	+	-	285°	+	-	+	285°
-	+	+	315°	-	+	+	315°
+	-	+	345°	+	+	-	345°

Mode 3: Half-step operation

- One-step and two-step modes are used alternately as shown in Fig.8.28(a).
- Excitation sequence: A, AB, B, BC, C, CA …. etc.
- The rotor revolves at a step angle of 15˚.

Phase-A excited

A1 and A2 are aligned with rotor teeth 1 and 3.
The rotor is now at an angle of $\theta = 0^\circ$.

Phase-A & B excited

Phase A and B are excited.
The rotor rotates through an angle of $\theta = 15^\circ$.
Clockwise direction.

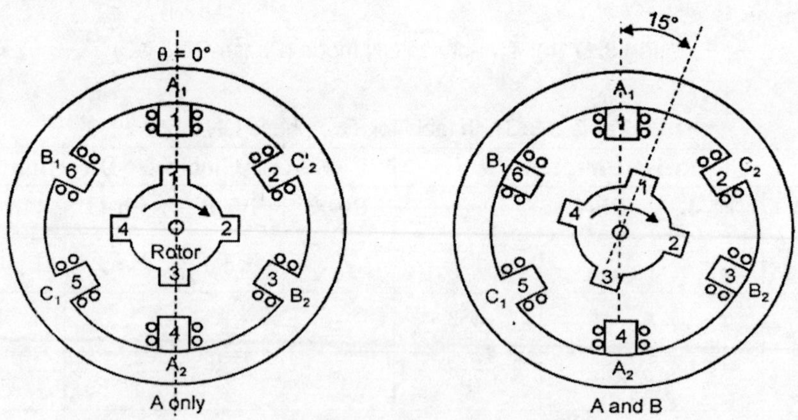

Fig.8.28 (a): Half step mode (Phase-A & AB)

Phase- B excited

- Phase B is the only one that is energised, whereas Phase A is de-energized, as shown in Fig.8.28(b).

- The rotor rotates through an angle of 15°.
- Rotor total movement is $\theta = 30°$
- Clockwise direction.

Phase- B & C excited

- Phases B and C are excited.
- The rotor rotates through an angle of 15°.
- Rotor total movement is $\theta = 45°$
- Clockwise direction.
- Stator sequence: A, AB, B, BC, C, CA – clockwise direction.
- Stator sequence: A, AC, C, CB, B, BA – anti-clockwise direction

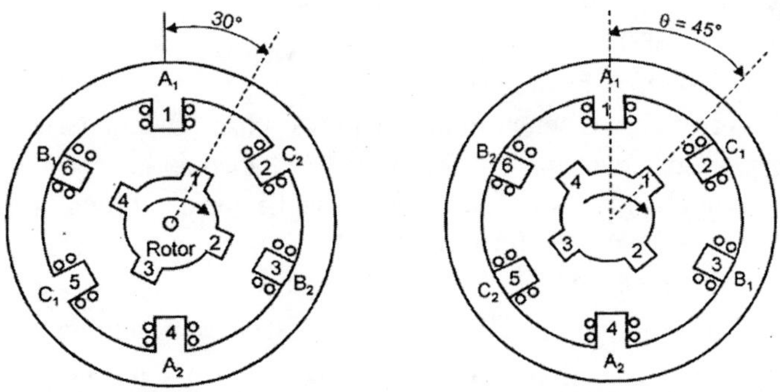

B only

BC winding only

Fig.8.28 (b): Half step mode (Phase-B & BC)

Table 8.4: Truth table Half-step operation

Clockwise Direction				Anti-Clockwise Direction			
Ph-A	**Ph-B**	**Ph-C**	**Degree**	**Ph-A**	**Ph-B**	**Ph-C**	**Degree**
+	-	-	0°	+	-	-	0°
+	+	-	15°	+	-	+	15°
-	+	-	30°	-	-	+	30°
-	+	+	45°	-	+	+	45°
-	-	+	60°	-	+	-	60°
+	-	+	75°	+	+	-	75°
+	-	-	90°	+	-	-	90°
+	+	-	105°	+	-	+	105°

Clockwise Direction				Anti-Clockwise Direction			
Ph-A	Ph-B	Ph-C	Degree	Ph-A	Ph-B	Ph-C	Degree
-	+	-	120°	-	-	+	120°
-	+	+	135°	-	+	+	135°
-	-	+	150°	-	+	-	150°
+	-	+	165°	+	+	-	165°

Mode 4: Micro stepping

- The step angle of the VR stepper motor is very tiny.
- Also known as small stepping.
- It employs two phases concurrently, as in two-phase ON mode.
- However, the two currents have been rendered unequal.
- Current through Phase A is held constant, while current through Phase B is gradually increased until the maximum current is attained.
- Phase-A is then reduced in incremental increments to zero.

When, $\beta=1.8^0$, Resolution $=360^0/\beta=360^0/1.8^0=200$ steps/revolution.

When, $\beta=0.0018^0$, Resolution $=360^0/\beta=360^0/0.0018^0=20{,}000$ steps/revolution.

- By employing micro stepping, fine resolution can be obtained.
- High resolution and smooth low-speed functioning.
- Used in printing, photo type setting etc.

ii. Multi-stack VR stepper motor

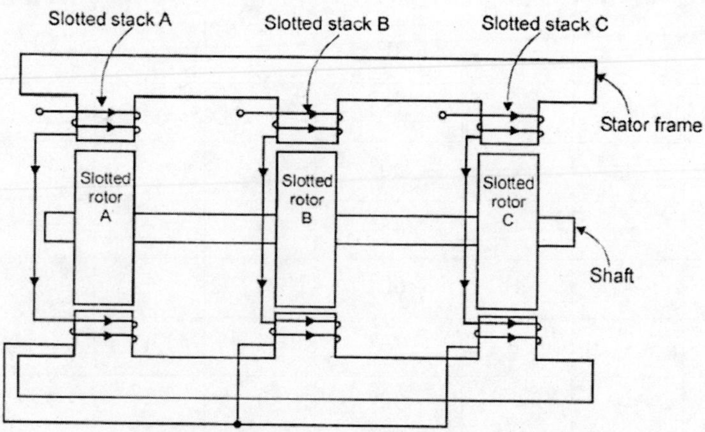

Fig.8.29: Stack Stepper Motor

The schematic of Multi-stack VR stepper motor is shown in Fig.8.29.

- Used to get a lower step size in the range of 2 to 15°.
- Stacks range from 3 to 7.
- Three-stack machines are quite prevalent.

Tooth pitch=$360^0/N_r=360^0/12=30^0$

Step angle = $360^0/mN_r=360^0/3 \times 12=10^0$

Each rotor pole is displaced by 10°

Operation

The phase excitation and the operation of 3-stack stepper motor is shown in Fig.8.30.

Phase-A excited

- Stack –A gets excited.
- The rotor poles of stack-A align with the stator poles.
- The rotor poles of stacks B and C are not aligned due to offset.

Phase-B excited

- Stack –B gets excited.
- The rotor poles of stack-B align with the stator poles.
- Rotor rotates by 10 degrees anticlockwise.
- The rotor poles of stacks A and C are not aligned due to offset.

Phase-C excited

- Stack –C gets excited.
- The rotor poles of stack-C align with the stator poles.
- Rotor moves by 10 degrees in an anti-clockwise manner.
- The rotor poles of stacks B and A are not aligned due to offset.

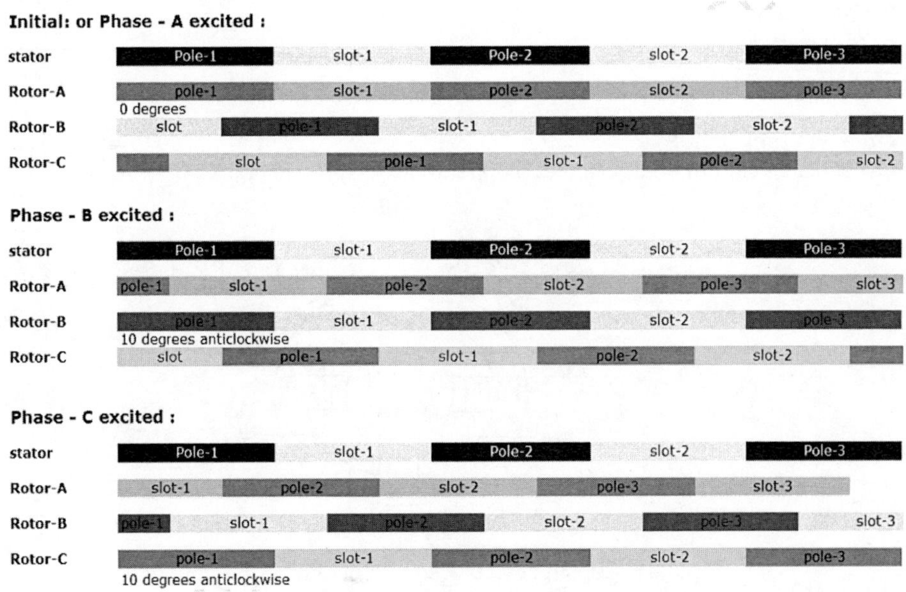

Fig.8.30: Operation of 3-stack stepper motor

Advantages

- Low rotor inertia
- A high torque/inertia ratio
- Capable of rapid stepping
- Fast slewing capability
- Light weight
- Three, four, and five phase, single and multi-stack variants are offered.
- Ability to freewheel

Disadvantages

- 3.6 degree step angles are often available.
- When the windings are de-energized, no detent torque is available.
- Shows mid-range resonance at particular stepping ratings under certain driving situations.
- Low efficiency at low voltages and steps per second.

B. Permanent Magnet Stepper Motor

A permanent magnet stepper motor's windings are subjected to phased alternating current. In practise, this is usually invariably a square wave produced by a direct current power supply. A square wave signal changing from plus to minus, for example, from +2.5 V to -2.5 V, is generated by the bipolar control system. The unipolar control system adjusts the direction of the magnetic flux of the coil by alternately supplying two signals to opposite terminals of the coil with respect to its central tap. There are several ways to control:

a. wave drive,
b. full step drive,
c. half step drive.

Construction

Construction of Permanent Magnet Stepper motor is given in Fig.8.31.

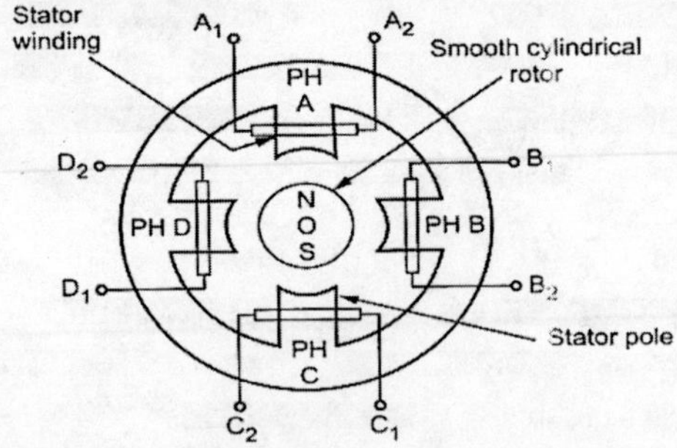

Fig.8.31: Construction of Permanent Magnet Stepper motor

Stator

- Stack of steel laminations

- Projected poles
- Multi polar type

Rotor

- Permanent magnet.
- Salient pole or cylindrical type.
- Cylindrical type used mostly.
- Made of ferrite or rare earth material.

Operation:

1-Phase ON mode

- One of the stator winding is energized.
- The rotor poles align with the energised stator poles.
- Stator windings can be energised with either current polarity.
- A+ Flow of positive current through phase winding A.
- A- Flow of negative current through phase winding A.

Phase A with + current

- Positive current is applied to phase winding A.
- The angle of rotation is zero. i.e., θ = 0°.

Phase B with + current

- Positive current is applied to phase winding B.
- Rotor rotates 90 degrees clockwise. i.e., θ = 90°.

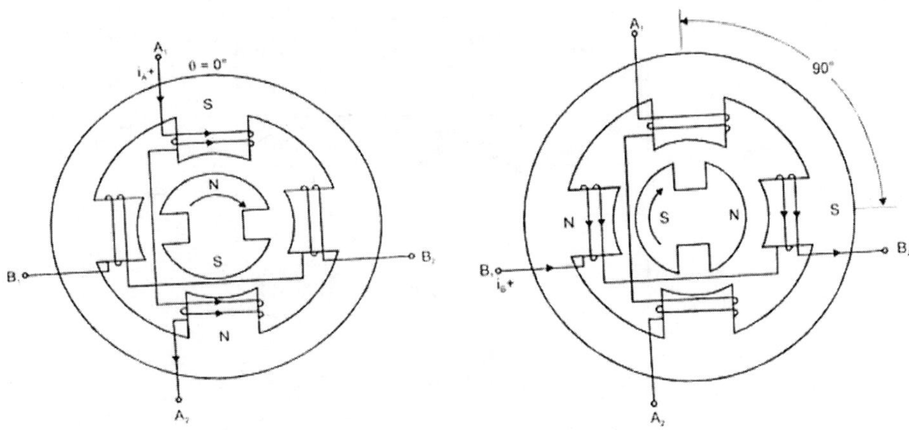

Fig.8.32 (a): Phase A & B with + current

Phase A with - current:

- Negative current is applied to phase winding A (ie) current flows from A2 to A1.
- The rotor revolves through 90° in a clockwise motion. θ = 180°.

Phase B with – current:

- Negative current is applied to phase winding B (ie) current flows from B2 to B1.
- The rotor revolves through 90⁰ in a clockwise motion. θ = 270⁰

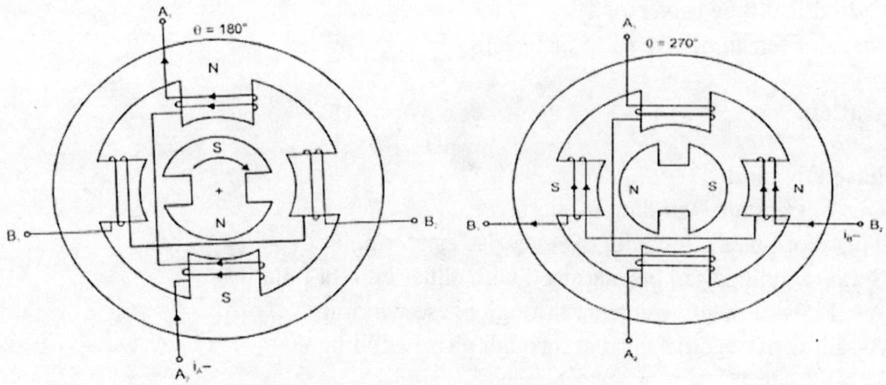

Fig.8.32 (b): Phase A& B with – current

- The rotor is then rotated through one complete 360-degree revolution after the phase winding A is energised.

Table 8.5: Truth table

Phase A	Phase B	Rotation θ
+	0	0^0
0	+	90^0
-	0	180^0
0	-	270^0
+	0	0^0

- The polarity of the phase currents determines the motor direction

Clockwise direction	I_A^+, I_A^+,I_A^-,I_A^-,I_A^+
Counter clockwise direction	I_A^+, I_A^-, I_A^-, I_A^+, I_A^+

C. Hybrid Stepper Motor

A hybrid stepper motor was created by combining the best characteristics of both stepper motors: variable reluctance and the permanent magnet, resulting in a reduced step angle. The rotor of the hybrid stepper motor is a cylindrical permanent magnet magnetised along the longitudinal axis with soft magnetic radial teeth. Typically, the stator has two or four phases that are spread between pairs of prominent poles. For the unipolar drive, the stator windings can be tapped centrally. Bifilar coiling is used for centre tap winding.

The hybrid stepper motor has:

- smaller step than a variable reluctance stepper motor or a permanent magnet stepper motor;
- the rotor is a fine-toothed permanent magnet. To lower pitch, the rotor's north and south teeth are offset by half of the tooth pitch;
- the tooth pitch of the stator poles is the same as that of the rotor;
- There are at least 2 phases in the stator;
- To provide a smaller step, the teeth of adjacent stator poles are offset by one-fourth of the tooth pitch.
- Combines VR and PM motors.

Rotor

- The rotor is made out of a permanent magnet with North and South poles at either end.
- End caps are also affixed at either end.
- End caps are Ferro-magnetic in nature.
- A half toothed pitch causes the two ends caps to be misaligned with respect to one another.
- At the rotor, they establish the alternate N and S poles.
- North poles are shown in full lines at the front end, while south poles are shown in dotted lines at the far end.

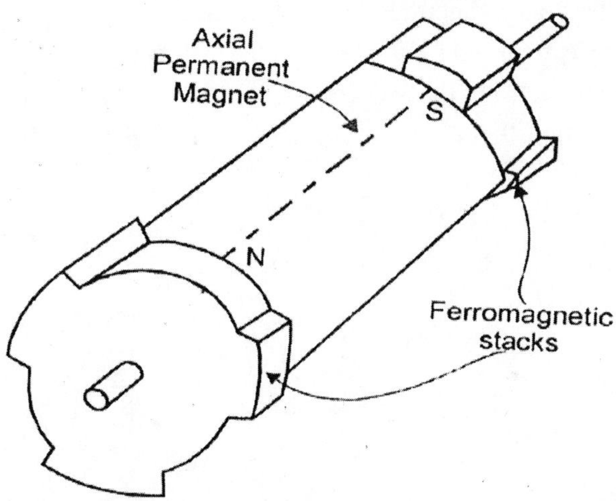

Fig.8.33: Construction of Hybrid Stepper motor

Operation:

- Stator has four phases: A_1,A_2, B_1, B_2.
- There are two phases: phase A and phase B.

Phase-A is energised by a positive current.:

- Pole A_1– north pole
- A_2– south pole.
- A1 attracts the extreme South Pole.
- A_2 attracts the front end north pole
- Rotor rotation is equal to zero degrees.

Phase-B is energised by a positive current:

- Pole B_1–north pole,
- B_2–south pole.
- B1 attracts the extreme South Pole.
- B_2 attracts the front end north pole
- Rotor rotation is equal to thirty degrees.
- Rotor total rotation, θ =30°
- Clockwise direction.

Phase-A energised with negative current:

- PoleA_1– south pole
- A_2–Northpole.
- A1 attracts the front end north Pole
- A_2 attracts the extreme end south pole
- Rotor rotation is equal to thirty degrees.
- Rotor total rotation, θ =60°

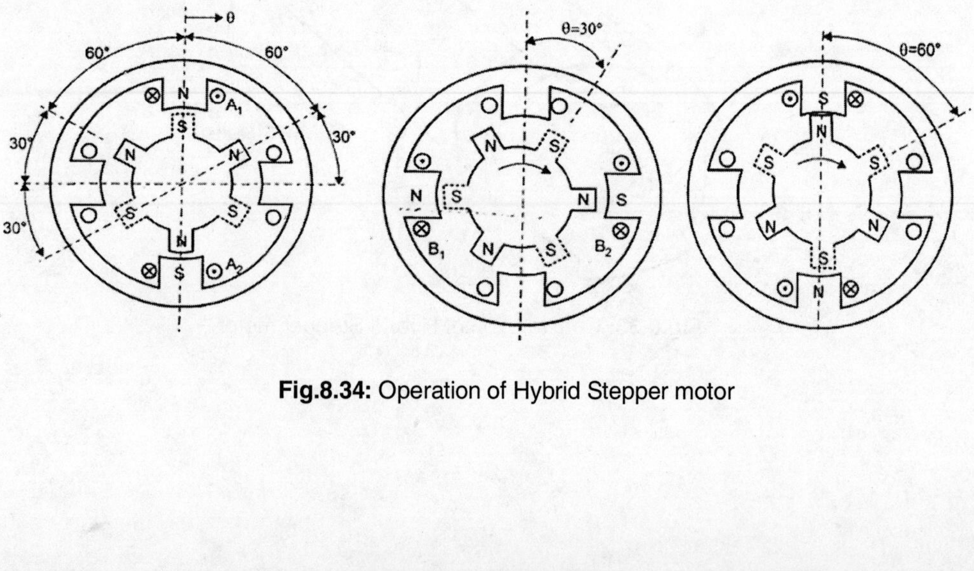

Fig.8.34: Operation of Hybrid Stepper motor

- A_1 B_1 A_2 B_2 A_1-clockwisedirection.
- $A_1B_2A_2B_1A_1$– counter-clockwise direction.

Advantages of hybrid stepper motor

- Provide détente torque with de-energized windings.
- Fewer resonance tendencies.
- Capability of handling high stepping rates.
- High efficiency at lower speeds and steps per minute.
- Increased holding torque capability.
- Improved damping due to the presence of a rotor magnet.

Disadvantages of hybrid stepper motor:

- Increased inertia and weight due to the presence of a rotor magnet.
- Changes in magnet strength have an effect on performance.

Distinctive features:

- Soft magnetic rotor with salient poles;
- The simplest and least expensive stepper motor;
- No detent torque in de-energized state;
- Large step angle.
- Permanent Magnet Stepper Motor

A permanent magnet rotor is used in a permanent magnet stepper motor. Typically, the stator has 2 phases. Stepper motors with an active rotor generate higher torques than variable reluctance motors and ensure the rotor remains fixed when the control signal is deleted. The downside of active rotor motors is the huge angular step (7.5-90 degrees). This is because manufacturing a permanent magnet rotor with a large number of poles is technically difficult. If the detent angle is between 7.5 and 90 degrees, the stepper motor is most likely a permanent magnet stepper rather than a hybrid stepper.

For use with a unipolar control circuit, the windings could be centred tapped. To power windings without a centre tap, bipolar control is required.

Thus, by the type of windings, two types of stepper motors are distinguished:

a. unipolar,
b. bipolar.

The hybrid stepper motor has

- The step is less than that of a variable reluctance stepper motor and a permanent magnet stepper motor;
- The rotor is a permanent magnet with fine teeth. The north and south teeth of the rotor are offset by half of the tooth pitch to reduce the pitch;

- The stator poles have the same tooth pitch as the rotor;
- The stator has at least two phases;
- The teeth of adjacent stator poles are offset by a quarter of the tooth pitch to create a smaller step.

8.6 AC and DC Servomotors

A DC motor that does not operate continuously for a longer period of time is what a servo motor is in essence. It features a special design that enables the motor to revolve with more accuracy and precision at a certain angle. A feedback mechanism is used to control this machine. Control motors are another name for servo motors. They do not employ continuous energy conversion; instead, they are used as output actuators in feedback control systems. Although the Servomotor's design and operation are distinct, its operating principle is the same as that of other electromagnetic motors. Their power rating ranges from a few hundred watts to a mere fraction of a watt.

The motors' rotor inertia is minimal and their response time is quick. The motor's rotor is narrower in diameter and longer in length. They work at extremely low speeds, and occasionally even zero speeds. Radar, computers, robotics, machine tools, tracking and guiding systems, processing controls, etc. all often employ servo motors.

8.6.1 Classification of Servo Motor

They are classified as

- AC Servo Motor
- DC Servo Motor

The AC servomotor is further divided into two types:

- Two Phase AC Servo Motor
- Three Phase AC Servo Motor

8.6.2 DC Servo Motor

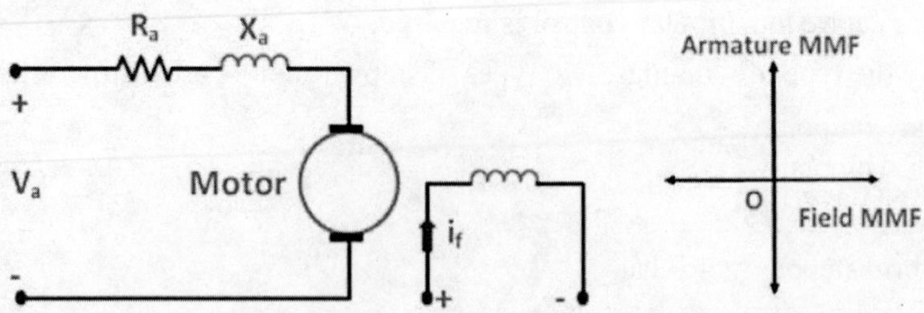

Fig.(a) Fig.(b)

Fig.8.35: (a) Connection of Separately Excited DC Servo motor
(b) Armature MMF and the excitation field MMF in quadrature

Permanent magnet DC motors or independently stimulated DC motors are both types of DC servo motors. The connection of a separately excited DC servo motor is shown in Fig.8.35 (a), and a DC machine's armature MMF and excitation field MMF are shown in quadrature in Fig.8.35 (b). Due to the decoupling of torque and flux, this offers a quick torque response. As a result, even a slight variation in the armature voltage or current causes the rotor's position or speed to alter considerably. The majority of high-power servo motors are mostly DC. The motor torque-speed characteristics is shown in Fig. 8.36.

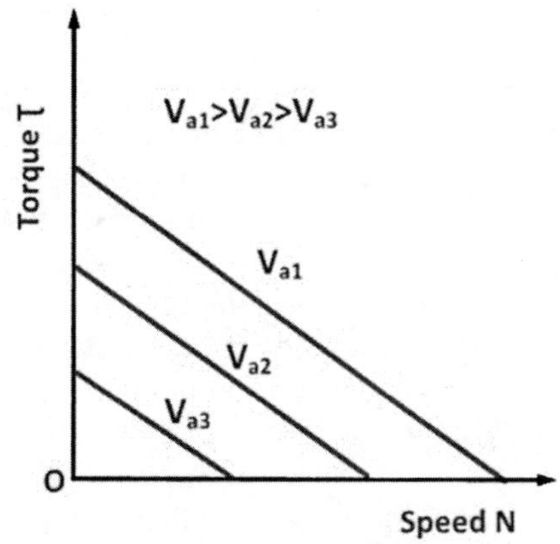

Fig.8.36: Torque-Speed Characteristics of DC Servo Motor

It is clear that the slope is negative based on the previously mentioned criteria. As a result, a negative slope gives the servo drive system viscous damping.

Applications

The servo motor has a wide range of uses because of its accurate and precise control, some of which are described below:

- Automation Industry
- Computers and robotics Industry
- Aviation
- Manufacturing Industry
- Pharmacy
- Food Services
- Tracking and guidance systems.

DC Servo Motor Advantages and Disadvantages

The benefits of a DC Servo Motor include

- Precise Control and Accuracy
- Stable Operation
- Fast Response

- Lightweight and portable
- Four Quadrant operation possible

The drawbacks of a DC servo motor include

- Due to the complex circuit, reliability is less
- Due to closed-loop components i.e. amplifier, gearbox, etc. it is costly

8.6.3. AC Servo Motor

The two varieties of AC Servo Motors are 2 and 3 Phase AC Servomotors. Two-phase squirrel cage induction motors are the most common form of AC servomotor. For low-power applications, they are employed. Induction motors with three phases are now used in applications that call for high power systems. An AC servo motor is a particular kind of servomotor that produces mechanical output utilising AC electrical input in the exact form of angular velocity. This servomotor's output power typically falls between watts and a few hundred watts. The range of an AC servo motor's working frequency is 50 to 400 Hz. Below is a schematic of an AC servo motor. The primary benefits of ac servo motors are their reduced weight, stability, and dependability during operation, lack of noise during operation, linear torque-speed characteristics, and lower maintenance costs in the absence of slip rings and brushes.

AC Servo Motor Construction

A two-phase induction motor is typically what an AC servo motor is. Similar to a typical induction motor, this one is built utilising a stator and rotor. This servo motor's stator typically has a laminated construction. As seen in Fig.8.6.3, this stator has two windings that are spaced 90 degrees apart. This phase variation causes a rotational magnetic field to be produced.

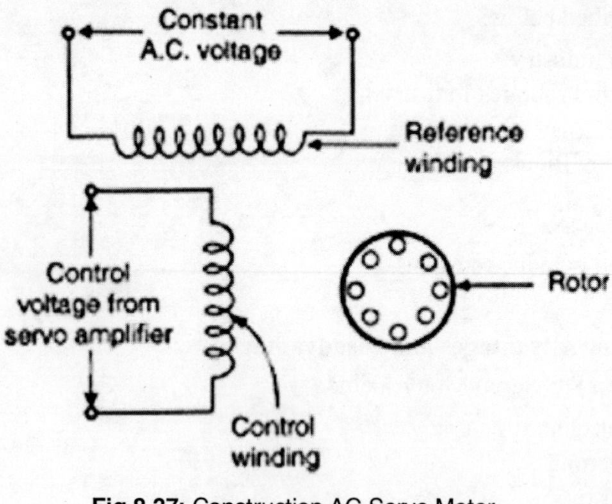

Fig.8.37: Construction AC Servo Motor

The first winding is known as the main winding or also known as fixed phase or reference winding. Here, the control winding or control phase is triggered by a changeable control voltage, whereas the main winding is operated by a source of constant voltage supply. A servo amplifier only supplies this control voltage. Generally, the rotor is available in two types of squirrel cage type & drag cup type. The rotor used in this motor is a normal cage-type rotor including using end rings as short-circuits and aluminium bars mounted in slots. For maximal flux linking, the air gap is kept to a minimum. The other type of rotor like a drag cup is mainly used where the inertia of the rotating system turns low. So this helps in decreasing power consumption.

Working Principle of AC Servomotor

The working principle of the ac servo motor is; firstly, a constant ac voltage is given at the startor's main winding of as seen in Fig.8.38, the servomotor and another stator terminal are simply linked to the control transformer through the control winding. Because of the applied reference voltage, the synchronous generator's shaft will rotate at a specific speed & obtains a certain angular position.

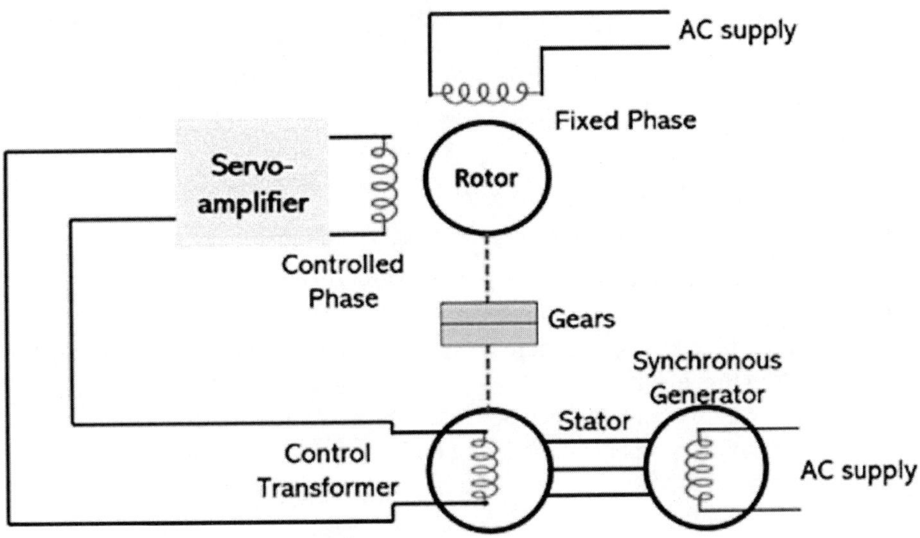

Fig.8.38: AC Servo Motor Circuit

In addition the shaft of the control transformer is compared to the shaft of the synchro generator at a certain angular point. Therefore, the error signal will be provided by comparing the two angular locations. More particularly, the levels of voltage for the equivalent shaft positions are evaluated which produces the error signal. So this error signal communicates with the present voltage level at the control transformer. After that, this signal is given to the servo amplifier so that it generates uneven control voltage. By this applied voltage, again In order to obtain the desired position of the motor inside the AC servomotors, the rotor reaches a specified speed, starts revolution, and continues until the error signal value reaches zero.

Characteristics of AC Servo Motor

Figure 8.6.5 depicts the torque speed characteristics of an AC servo motor. Due to its primary dependence on the ratio of reactance (X) to resistance (R), the torque in the following characteristics changes with speed but not linearly. When this ratio is low, the motor's properties are more linear than when it is high, which indicates that the motor has high resistance and low reactance.

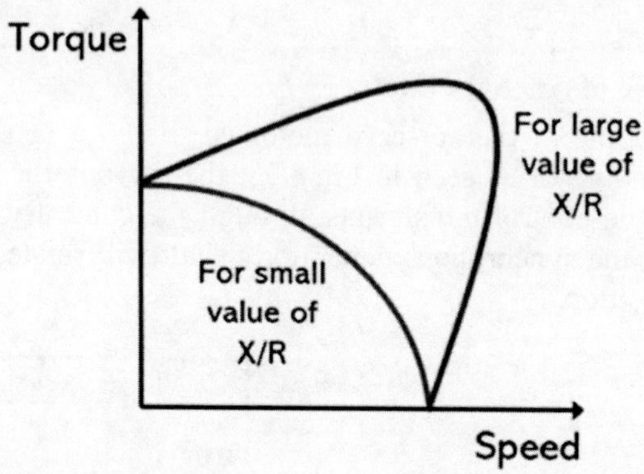

Fig.8.39: Torque-Speed Characteristics of AC Servo Motor

Advantages and Disadvantages

The following are some benefits of using AC servo motors.

- The speed control characteristics of this motor are good.
- They generate less amount of heat.
- They offer high efficiency, more torque per weight, reliability & reduced RFnoise.
- They need less maintenance.
- They have a longer life expectancy in the nonexistence of a commutator.
- These motors are capable of handling higher current surges in industrial machinery.
- At high speeds, they offer more constant torque.
- These are highly reliable.
- They provide high-speed performance.
- These are well-suitable to unstable load applications.

The disadvantages of AC servo motors include the following.

- AC servo motor control is more difficult.
- These motors can be broken by constant overload.
- Gearboxes are frequently necessary to transmit power at high speeds.

Applications

The applications of AC servo motors include the following.

- AC servo motors are useful in situations where position control is important and are frequently used in machine tools, robotics, and semiconductor devices.

- These motors are used in the instruments which operate on servomechanism like in computers & position control devices.
- AC servo motor is used in machine tools, robotics machinery & tracking systems.
- These servo motors are used in a variety of industries because of their efficiency& versatility.
- The AC servo motor is used in most common machines & appliances like water heaters, ovens, pumps, Off-road vehicles, equipment in gardens, etc.
- Many of the appliances & tools that are used every day around the house are power-driven by AC servo motors.

Comparison between AC servomotor and DC servomotor

The following table compares AC servomotor and DC servomotor

Table 8.6: Comparison between AC and DC servomotor

S. No.	AC servo motors	DC servo motors
1.	Low power output	High power output
2.	Less efficiency	High efficiency
3.	Less maintenance due to absence of commutator	Frequent maintenance due to commutator
4.	Less stability problem	More stability problem
5.	No radio frequency noise	More radio frequency produces due to brushes
6.	Stable and Smooth operation	Noisy operation

Comparison between Stepper and Servo Motor

The following table compares Stepper and Servo Motor

Table 8.7: Comparison between Stepper and Servo Motor

S. No	Basis for Comparison	Stepper Motor	Servo Motor
1.	Basic	Stepper motor operates insteps.	It is continuous operating machine.
2.	System configuration	Open loop	Closed loop
3.	Power requirement	More	Comparatively less
4.	Design	Simple	Complex

S. No	Basis for Comparison	Stepper Motor	Servo Motor
5.	Ability to response	High	Comparatively low
6.	Cost	Inexpensive	Expensive
7.	Reliability	More	Less
8.	Noise and vibration	High	Comparatively less
9.	Operating speed	Slow	Fast
10.	Feedback mechanism	Not exist	Exist
11.	Heat generation	More	Comparatively less
12.	Number of poles	Generally 50 to 150	Around 4 to 12
13.	Life span	More	Less
14.	Damage due to overload	Less prone to get damaged.	Comparatively more prone to get damaged.
18.	Torque produced	High	Low
16.	Efficiency	Less	More
17.	Tolerance towards moment of inertia	High	Low
18.	Applications	In 3D printing equipment, games, textiles, welding equipment, and other industries.	Systems for placing antennas, automated doors, cameras, remote controlled devices, etc. in robotics.

Difference between Servo Motor and DC Motor

The following table compares Servo Motor with DC Motor in tabular form.

Table 8.8: Comparison between Servo Motor and DC Motor

S. No	BASIC	SERVO MOTOR	DC MOTOR
1.	Wire system	Power, ground, and control are the three wires that make up a servo motor.	Power and ground are the two wires that make up a DC motor.
2.	Assembly	It consists of a DC motor assembly, a gearing set, a control circuit, and a position sensor.	A DC motor is a standalone device that requires no assembly.

S. No	BASIC	SERVO MOTOR	DC MOTOR
3.	Rotation	Unlike a DC motor, a servo motor does not rotate constantly and freely. It only rotates 1800 times per minute.	A DC motor moves continuously.
4.	Examples	They are utilised in robotic rudder control, arms, and legs.	Car wheels, fans, and other devices employ DC motors.

UNIT SUMMARY

This comprehensive unit provides a holistic understanding of different types of fractional motors like Synchronous Reluctance Motor, Switched Reluctance Motor, BLDC, Permanent Magnet Synchronous Motors, Stepper motors, AC and DC Servomotors. Construction, working Torque speed characteristics and applications of above motors were discussed.

EXERCISES

1. A brushless DC motor has 1212 slots on the stator and 1010 magnets on the rotor. If the motor operates at 15001500 rpm, calculate the electrical and mechanical angles per pole.

2. Draw torque-speed characteristics for a Permanent Magnet Synchronous Motor and compare them with a Synchronous Reluctance Motor for a given load torque.

3. Identify and explain the application scenarios where a Switched Reluctance Motor is preferred over a BLDC motor.

Multiple Choice Questions

1. Which of the following is/are the main advantages of synchronous reluctance motor?

   a. Cheaper

   b. Higher efficiency

   c. Robust

   d. All of the above

2. Which one of the following is an application of synchronous reluctance motor?

   a. Pumps or conveyers

   b. Nuclear reactors

   c. Both a and b

   d. None of the above

3. Which one of the following is a useful property of synchronous reluctance motor?
   a. High inductance
   b. High speed capability
   c. High temperature capability
   d. All of the above

4. The figure shown below is a construction of ______ motor?
   a. Induction
   b. Reluctance
   c. Brushless
   d. None of the above

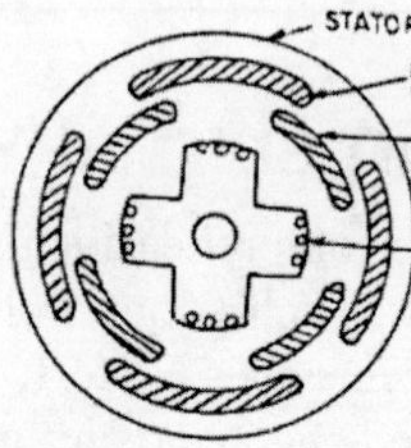

5. Which one of the following motor costs is high?
   a. Synchronous reluctance motor
   b. Induction motor
   c. Both a and b
   d. None of the above

6. Which of the following is employed to enhance the operating for a switched type reluctance motor?
   a. Switching diode
   b. Switching inverter
   c. Switching transistor
   d. All the above

7. In a switched type reluctance motor, which of the following factors affects magnetic type circuit?
   a. Air gap
   b. Short circuit
   c. Open circuit
   d. All the above

8. A switch-type reluctance motor is designed using which of the following materials?
   a. Si Steel
   b. Al steel
   c. Pt steel
   d. Steel

9. The field winding in a BLDC motor is kept up on ______________
   a. Stator
   b. Rotor
   c. Can be placed anywhere
   d. Absent

10. Which of the following is not a benefit of a BLDC motor over a traditional DC motor?
   a. Less maintenance
   b. Long life
   c. No risk of explosion or possibility of RF radiation
   d. Low cost

11. Which DC motor is frequently used in medical fields?
   a. PMDC
   b. BLDC
   c. Brushed DC motor
   d. Cannot be determined

12. BLDC construction is same to that of ______________
   a. Conventional DC motor
   b. Induction motor
   c. Permanent magnet synchronous motor
   d. Totally different construction

13. A typical brushless motor lacks ______________
   a. Commutator
   b. Permanent magnet
   c. Electronic controller
   d Fixed armature

14. Alternatively, BLDC can be used for ____________
   a. Synchronous motor
   b. Normal brushed DC motor
   c. Induction motor
   d. Air motor

15. The number of windings on the field of permanent magnet d.c. motor is
   a. More compared to shunt motor
   b. Less compared to shunt motor
   c. Absent
   d. None of these

16. Permanent magnet d.c. motors are extensively used in
    a. Automobiles
    b. Heaters
    c. Air conditioners
    d. All of these

17. Which of these motors revolves in discrete angular steps?
    a. Servo motor
    b. DC motor
    c. Stepper motor
    d. Linear Induction Motor (LIM)

18. The stepper motor responds to
    a. A programmed sequence of input electrical pulses.
    b. Pulse Width Modulation (PWM).
    c. feedback signal.
    d. Pulse Position Modulation (PPM).

19. A variable reluctance stepper motor's rotor is made of ______material and has salient poles.
    a. Ferromagnetic
    b. Diamagnetic
    c. Non-magnetic
    d. Paramagnetic

20. A variable reluctance stepper motor has eight primary poles, each with five teeth. Calculate the stepping angle for a rotor with 60 teeth.
    a. 0.9 degree
    b. 3 degree
    c. 0.5 degree
    d. 1.8 degree

21. The step angle of a stepper motor is 2.50. Determine the number of steps required for the shaft to turn twenty rotations
    a. 3600
    b. 2500
    c. 144
    d. cannot be determined

22. The stator and rotor have the same number of poles and thus the same pole pitch.
    a. Permanent Magnet stepper motor
    b. Single stack variable reluctance stepper motor
    c. Multi stack variable reluctance stepper motor
    d. Hybrid stepper motor

23. The combines the characteristics of the Variable Reluctance Stepper Motor and the Permanent Magnet Stepper Motor.
    a. Multi stack variable reluctance stepper motor
    b. Hybrid stepper motor
    c. Servo motor
    d. None of the above

24. The rotational speed of a stepper motor is dependent on
    a. Magnitude of supply voltage.
    b. Polarity of stator current.
    c. Magnitude of stator current.
    d. Step pulse frequency.

25. Which of the following is most accurate motor?
    a. Squirrel cage induction motor
    b. Universal motor
    c. Servomotor
    d. Repulsion motor

26. A DC servo motor consists of
    i. a small DC motor,
    ii. feedback circuit,
    iii. gearbox.
    a. Only (i)
    b. Only (ii) and (iii)
    c. All of the above
    d. only (i) and (ii)

27. Consider the following statements regarding DC servomotor:
    i. The speed of servomotor can be controlled by varying the armature voltage.
    ii. The armature is deliberately designed to have large armature resistance.
    a. Only statement (i) is correct.
    b. Only statement (ii) is correct.
    c. Both statements (i) and (ii) are correct.
    d. Both statements are wrong.

28. In AC servomotor main winding and control winding are displaced by
    a. 90° electrically
    b. 90° mechanically
    c. 180° electrically
    d. 180° mechanically

29. One of the basic requirements of servomotor is that it must produce high torque at all
    a. voltages
    b. loads
    c. speeds
    d. frequencies

30. DC servomotors are used in
    a. Purely AC control systems
    b. Both AC & DC control systems
    c. Purely DC control systems
    d. None of these

Answers of Multiple-Choice Questions

1 d. All of the above
2 c. Both a and b
3 d. All of the above
4 b. Reluctance
5 a. Synchronous reluctance motor
6 c. Switching Transistor
7 a. Air Gap
8 a. Si Steel
9 b. Rotor
10 d. low cost
11 b. BLDC
12 c. Permanent Magnet Synchronous Motor
13 a. Commutator
14 b. Normal Brushed DC motor
15 c. Absent
16 d. All of these
17 c. Stepper Motor
18 a. a programmed sequences of input electrical pulses
19 a. Ferromagnetic
20 a. 0.9 degree
21 a. 3600
22 c. Multi stack variable reluctance stepper motor
23 a. Multi stack variable reluctance stepper motor
24 d. Step pulse frequency.
25 c. Servomotor
26 c. All of the above
27 c. Both statements (i) and (ii) are correct.
28 a. 90^{0} electrically
29 b. loads

30 c. Purely DC control systems

Short and Long Answer Type Questions

Long Answer Questions with Answers

1. Compare synchronous reluctance motor and induction motor?

S. No.	Synchronous Reluctance Motor	Induction Motor
1	Better efficiency	Poor efficiency
2	High cost	Low cost
3	Low power factor	High power factor
4	Used for low and medium power applications.	Used for high power applications.

2. Mention some advantages and disadvantages of synchronous reluctance Motor.

Advantages:

- There is no demagnetization, hence synchronous reluctance.
- There need be no excitation field at zero torque, thus eliminating electromagnetic spinning losses.
- Rotors can be constructed entirely from high strength, low-cost materials.

Disadvantages:

- Compared to induction motor it is slightly heavier and has low power factor. But increasing the saliency ratio, the power factor can be improved.
- High cost than induction motor.
- Need speed synchronization to inverter output frequency by using rotor position sensor and sensor less control.

3. Write at least five properties of synchronous reluctance motor.

- High output power capability.
- Ability of the rotor to withstand high speeds.
- Negligible zero-torque spinning losses.
- High reliability.
- Lower torque ripple

4. What are the design considerations in synchronous reluctance motor?

The design considerations of synchronous reluctance motor are,

- Power factor
- Copper loss and core loss
- Cost
- Efficiency

5. Give the operating principle of radial flux motor.

In radial flux motor, the magnets are alternately poled and radially magnetized, but because the magnet pole area is smaller than the pole area at the rotor surface, the air-gap flux density on open circuit is less than the flux density in the magnet; this design is essentially 'under excited' and relies on the addition of a magnetizing component of armature current to produce the total air gap flux.

6. Compare synchronous reluctance motor and Switched reluctance motor.

S. No.	Synchronous reluctance motor	Switched reluctance motor
1.	Equal no. of stator and rotor poles.	It is a form of stepper motor that has lesser fewer poles.
2.	Operate at pure sinusoidal voltage.	Optimal drive waveform is not a pure sinusoidal.
3.	The stator is cylindrical type with distributed winding.	The stator of SRM has salient poles with concentrated coils like a dc motor.
4.	Used for low and medium power applications.	Used for high power applications.
5.	Speed control requires a variable frequency drive.	Does not need any variable-frequency drive.
6.	Excitation is a set of poly phase balanced sine wave currents.	Excitation is a sequence of current pulses applied to each phase.

7. Distinguish between axial and radial air gap motors?

S. No	Axial air gap motors	Radial air gap motors
1	Axially laminated rotor.	Radially laminated rotor.

S. No	Axial air gap motors	Radial air gap motors
2	By increasing L_d / L_q ratio, we obtain more power factor and efficiency.	By decreasing L_d / L_q ratio, circulating flux in the rotor pole faces.
3	Designed to have high saliency.	Designed to have optimized flux guide.
4	Rotor has two design structures.	Rotor has one design structure.
5	Good choice for high speed applications.	Poor choice for high speed applications.

8. What benefits does a switching reluctance motor offer?

- Robust with simple construction.
- The Rotor requires minimal maintenance because it has no brushes, slip rings, or windings.
- A permanent magnet does not exist.
- The ventilation system is easier because losses primarily occur in the stator.
- Simpler power semiconductor switching circuitry
- Power short circuits are not likely to shoot through faults,
- The polarity of the current in the phase winding has no bearing on the amount of produced torque.
- By adjusting the region of conduction, the machine's operation can be quickly switched from driving mode to generating mode.
- Obtaining a very high speed is achievable.
- The self-starting device.
- Through the use of feedback diodes, energy stored in the phase winding is returned to the supply.

9. What are the benefits and drawbacks of stepper motors?

Benefits

- It can be driven in an open loop with no feedback.
- Because stepper motors respond immediately to digital control signals, they are a perfect candidate for digital computer controllers.
- It is mechanically straight forward.
- It requires less maintenance.

Drawbacks

- Low efficiency with standard controllers.
- Fixed step angle.
- Limited capacity to handle big inertia loads.
- Limited power output and size options.

10. Explain about how the SRM works.

A sequence of current pulses are applied at each phase stimulate the motor. As a result, individual phases are excited, pushing the motor to rotate. Rotor attempts to align in either the maximum inductance or minimum reluctance position. The stator phase receives electricity, which causes the motor to produce torque in the direction of increasing inductance.

Phase A energized: aa' gets aligned with AA'. Minimum reluctance position. Inductance is maximum.	**Phase B energized:** bb' gets aligned with BB'. Minimum reluctance position. Inductance is maximum.	**Phase C energized:** aa' gets aligned with CC'. Minimum reluctance position. Inductance is maximum.
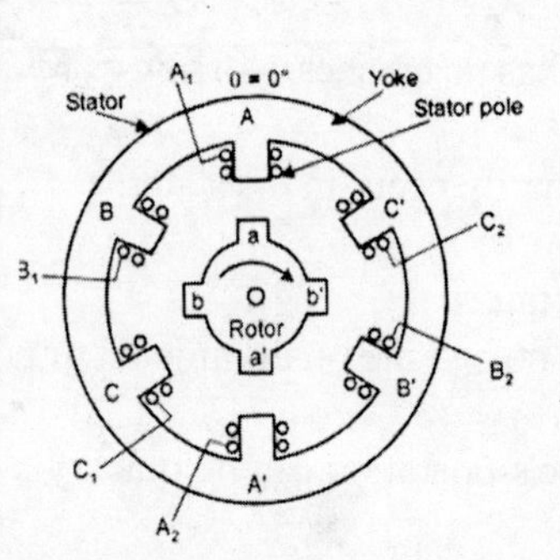	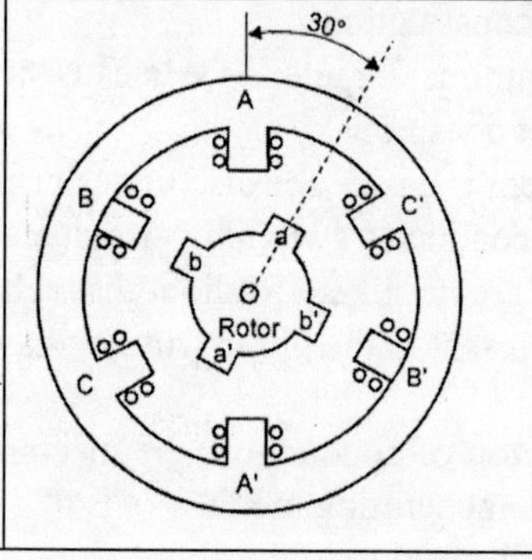	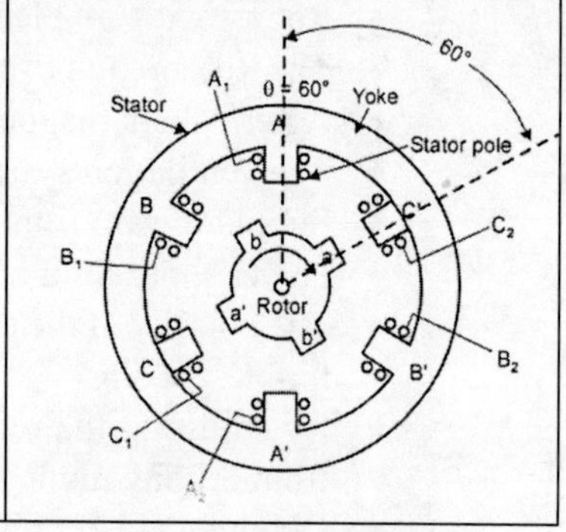

11. List out the advantages and disadvantages of the converter circuit with two power semiconductor devices and two diodes per phase.

Advantages:

- Reduces switching losses of converter circuit.
- Control of each phase is completely independent of the other phases.

Disadvantages:

- Converter circuit expensive

12. Mention the converter circuit's benefits and drawbacks. It has two power semiconductor devices and two diodes for each phase.

Advantages:

- Each phase's control is totally independent of the operation of the other phases,
- Lowers the converter circuit's switching losses.

Disadvantages:

- Expensive converter circuit

13. Differentiate between Conventional DC motor and PMBLDC motor.

Comparison between conventional DC motor and PMBLDC motor:

S. No.	Features	Conventional DC motor	PMBLDC motor
1.	Mechanical structure	On the stator, there are field magnets.	The rotor is equipped with field magnets.
2.	Maintenance	High Maintenance	Low maintenance
3.	Life	Shorter	Longer
4.	Speed/Torque characteristics	Moderately flattened - as speeds increase, brush friction increases, lowering useable torque.	Flat - Allows for operation at all speeds with the rated load.
5.	Efficiency	Low / Moderate	High
6.	Rotor inertia	High	Low
7.	Speed range	Low – Mechanical constraints	High – No mechanical constraints
8.	Efficiency	Moderate	High
9.	Control	Simple & inexpensive	Complex and expensive

14. Comparison between mechanical and electronic commutator.

S. No	Mechanical Commutator	Electronic commutator
1.	Commutator segments and mica insulation make up the commutator. Carbon graphite is used to make brushes.	The commutator employs power electrical switching components.
2.	The rotor houses the commutator assembly.	The stator houses the commutator assembly.
3.	The arrangement includes shaft position sensing.	It necessitates the use of a separate rotor position sensor.
4.	The number of commutator segments is extremely large.	The number of switching devices is restricted to six.
5.	Sliding Contact between the commutator and the brushes.	There are no sliding contacts.

S. No	Mechanical Commutator	Electronic commutator
6.	Sparking takes place.	There is no sparking.
7.	It must be serviced on a regular basis.	It takes very little maintenance.
8.	Controlling the voltage available across tappings is difficult.	PWM techniques can be used to regulate the voltage available across armature tapping.
9.	Exceptionally reliable at all.	Specially designed devices and protection circuits can improve reliability.

15. Explain the operation of different types of BLPM DC motor.

Types of BLDC based on magnetic arc:

The flux-density distribution in the motor's air gap can be used to classify BLPMDC motors. They do,

- BLPM square wave DC motor.
- BLPM sine wave DC motor.

16. What are the benefits and drawbacks of VR stepper motors?

Benefits:

- Low rotor inertia
- High torque to inertia ratio
- Light weight
- Capable of high stepping rate
- Ability to free wheel

Drawbacks:

- Normally available in 3.6° to 3^0 step angles.
- No detent torque available with windings de–energized.

17. What are the benefits and drawbacks of a permanent magnet stepper motor?

Benefits:

- Low power demand
- High detente torque as compared to VR motor
- Rotor do not require external exciting current
- It produces more torque per ampere stator current

Drawbacks:

- The motor has more inertia.
- Decreased acceleration

18. What are the benefits and drawbacks of a hybrid stepper motor?

Benefits:

- Less likely to resonance.
- Provide détente torque when windings are de-energized.
- Increased holding torque capability
- Capability for high stepping rates

Drawbacks:

- Increased inertia and weight due to the presence of a rotor magnet.
- Performance impacted by changes in magnetic strength.

19. Draw a stepper motor's usual static characteristics.

The Torque- Angle and Torque-Current characteristic of stepper motor are shown in figure below.

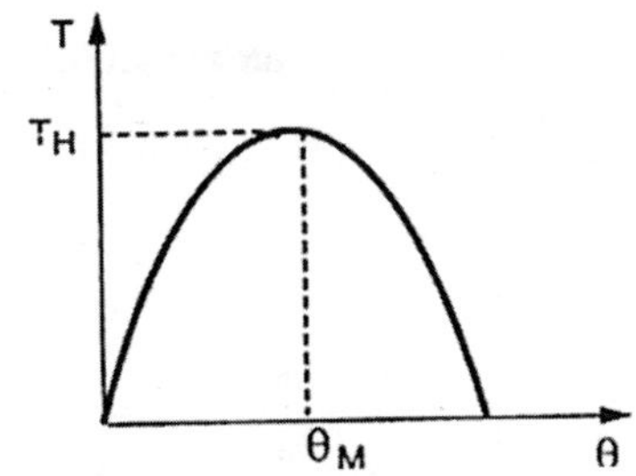

Torque- Angle Characteristic

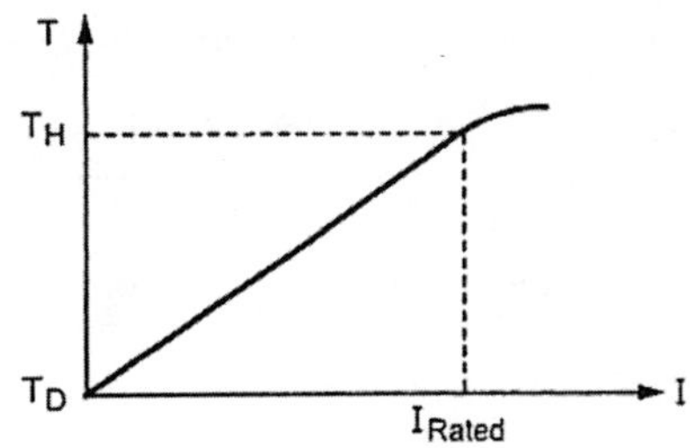

Torque- Current Characteristic

20. Distinguish between VR, PM, and hybrid stepper motors.

S. No.	VR Stepper motor	PM Stepper motor	Hybrid stepper motor
1	Less rotor inertia	More inertia	More inertia
2	Light weight	Heavy weight	Heavy weight
3	Detent torque not available	Detent torque is available	Detent torque is available with de-energized windings
4	No permanent magnet Rotor	Permanent Magnet Rotor	Permanent Magnet Rotor
5	Salient pole type Rotor	Cylindrical Rotor	Salient pole type Rotor

21. Differentiate between the half step and full step activities of a stepping motor.

S. No.	Half step operation	Full step operation
1.	The half step operation, also known as half stepping, can be obtained by ctivating the three phases in the sequence a, ab, b, bc, c, and so on.	The entire operation can be obtained by activating only one phase at a time.
2.	This mode, also known as wave excitation	This mode is also known as one-phase - ON mode.
3.	Produces smoother shaft rotation through continuous half stepping..	This is the most basic and generally used method of making the motor step.

22. In stepping motors, compare single stack and multi stack arrangements.

S. No.	Single stack	Multi stack
1.	The number of stator poles should be distinct from the number of rotor poles.	The number of stator poles should be equal to the number of rotor poles.
2.	A field coil is attached to each stator pole.	It is employed to achieve small step sizes. It is made up of m identical single stack variable reluctance motors with rotors mounted on a single shaft.

Short Questions with Answers

1. What is a synchronous reluctance motor?

It is the motor driven by reluctance torque which is produced due to tendency of the salient rotor poles to align themselves with synchronously rotating field produced by stator. In this motor, the magnets are left out of the rotor or they are demagnetized. The rotor of asynchronous reluctance motor has salient poles but neither have field windings nor permanent magnets.

2. What are the types of synchronous reluctance motor?

A. The main types are,
   i. Cageless
   ii. Line-start

B. According to the magnetization (when the stator winding is energized),
   i. Radial type
   ii. Axial type

3. Why are SR machines so common in variable-speed drives?

A common application for SR machines in adjustable speed drives is the control of driving a motor in both motoring mode and generating mode of operations.

4. What is the purpose of switching reluctance motor closed loop control?

The switching reluctance motor benefits from closed loop control because it has improved repeatability and stability.

5. State the principle of operation of synchronous reluctance motor.

When a piece of magnetic material is located in a magnetic field, a force acts on the material tending to bring it into the most dense portion of the field. The force tends to align the specimen of material in such a way that the reluctance of the magnetic path that lie through the material will be minimum. In general, reluctance torque is developed by the tendency of a ferromagnetic material to align itself with a magnetic field. (i. e.,) when the stator winding is energized, the evolving magnetic field produces reluctance torque.

6. State any four advantages of synchronous reluctance motors.

- It can operate from essentially standard P.W.M. a.c. inverters.
- Lower torque ripple.
- Simple and rugged construction.
- It has high speed capability.

7. List any four applications of synchronous reluctance motor.

- The synchronous reluctance motor is widely used for many constant speed applications such as recording instruments, timings devices, control apparatus and photograph.
- It is employed for low power application such as spinning mills.
- Used in processing of continuous sheet or film material;
- Applied in synthetic fibre manufacturing equipment.

8. Define Torque Angle.

In reluctance type synchronous motor, when the load is increased lightly, the rotor momentarily slows down, causing the salient poles of the rotor to lag the rotating field. This angle of lag is called the torque angle.

9. Define reluctance torque.

In a synchronous reluctance motor, the torque which is produced at critical speed, due to the tendency of the salient rotor poles to align themselves with synchronously rotating field produced by the stator is known as reluctance torque. (i.e.,) the reluctance torque is produced, when the low-reluctance path provided by the salient rotor poles causes them to snap into synchronism with the rotating flux of the stator and the rotor is pulled around by simple magnetic attraction.

10. Compare PMBLDC motor and switched reluctance motor?

S. No.	PMBLDC Motor	Switched Reluctance Motor
1.	A permanent magnet is a rotor.	There is no permanent magnet in the rotor.
2.	Price is high	When compared to PMBLDC, the cost is lower.
3.	Efficiency is high	Efficiency is less

11. How are the directions of rotations reversed in case of PMBLDC motor?

Altering the direction of a BLDC motor:

For Anticlockwise direction:	For clockwise direction:
PT1 – Tr1	PT1 – Tr3
PT2 – Tr2	PT2 – Tr1
PT3 – Tr3	PT3 – Tr2

Where PT1, PT2, PT3 – photo transitors.

12. What is meant by demagnetization in PMBLDC motor?

The second quadrant of the B-H loop is the most critical. This is referred to as the demagnetization curve. In the lack of externally provided ampere-turns, the magnet's operating point lies at the point of intersection of the demagnetization curve and the load line, the slope of which is the product of μ_0 and the external circuit's permeance coefficient.

13. How the demagnetization occurs in PMBLDC motor?

Demagnetization occurs during normal motor operation when the torque and back emf are constant and the field flux level gets low.

14. When does a PM synchronous motor operate as synchronous reluctance motor?

If the magnets are removed or demagnetized, the permanent magnet synchronous motor operates as a synchronous reluctance motor.

15. What are the differences between conventional synchronous motor and PM synchronous motor?

S. No	Conventional synchronous motor	PM synchronous motor
1.	Field winding of the rotor is provided by a DC supply through slip rings and brushes.	Magnets are permanent in the rotor.

S. No	Conventional synchronous motor	PM synchronous motor
2.	The presence of slip-rings and brushes increases the amount of maintenance required.	Easy to maintain.
3.	The DC field is maintained off when the wound field synchronous motor is started as an induction motor.	There is no way to "turn off" the field.
4	no sensor for rotor position.	Using a sensor, the location of the rotor shaft may be determined.

16. List out the differences between the PM brushless DC motors and PM synchronous motors.

S. No.	PM brushless DC motor (Square wave)	PM synchronous motor (Sine wave)
1	The stator's winding has been focused.	It has distributed winding on the stator.
2	Produces voltage that is square or trapezoidal.	Induces sinusoidal voltage.
3	Known as brushless DC motor.	Known as brushless AC motor.
4	A low-power drive is employed.	A high power drive is used.

17. What are the applications of a servo motor?

Automation, Robotic, Manufacturing, etc.

18. What is the definition of holding torque?

Holding torque is the maximum load torque that an energised stepper motor can withstand without falling out of balance.

19. What is the definition of detent torque?

Detent torque is the maximum load torque that an unenergized stepper motor can withstand without slipping. It is also known as cogging torque.

20. Describe some stepper motor applications?

i. Floppy disk drives.

ii. Quartz watches.

iii. Camera shutter operation.

iv. Dot matrix and line printers.
v. Machine tool applications.
vi. Robotics.

Short and Long Answer Type Questions

Long answer Questions

1. Describe the constructional features of synchronous reluctance motor.
2. What are the types of synchronous reluctance motor? With neat diagrams, explain the same.
3. Explain the operation of axial type synchronous reluctance motor.
4. Explain the operation of radial type synchronous reluctance motor.
5. Explain briefly working principle and operation of synchronous reluctance motor.
6. Draw and explain the steady state phasor diagram of synchronous reluctance motor
7. Draw and discuss a typical torque-speed characteristics of synchronous reluctance motor.
8. Enumerate the advantages, disadvantages and applications of synchronous reluctance motor.
9. Draw and explain torque-angle curve and speed - torque characteristics of synchronous reluctance motor.
10. Explain the Switched Reluctance Motor's (SRM) constructional characteristics and operation.
11. Explain how a microprocessor-based controller controls a switched resistance motor with a neat sketch.
12. Describe the Switched Reluctance motor's Speed-Torque characteristics.
13. State some applications of SRM?
14. What benefits does a switching reluctance motor offer?
15. Explain briefly the construction and working principle of PMBLDC motors.
16. Explain the operation of PMBLDC motor.
17. List out the differences between conventional DC motor and PMBLDC motor?
18. What are the advantages of BLPM dc motor over conventional DC motor?
19. State the principle of operation of PM brushless DC motor.
20. What is an electronic commutator?
21. List any four permanent magnet materials.
22. Compare PMBLDC motor and switched reluctance motor?
23. List out some applications of BLPMDC motor?
24. Comparison between mechanical and electronic commutator.
25. Explain briefly constructional and working principle of PMBLDC motors.
26. Explain the operation of PMBLDC motor.
27. Explain briefly about the mechanical commutator in conventional DC motor and electronic Commutator in BLPMDC motor.
28. Analyse the operation of electronic commutator in PMBLDC motor with neat diagram.
29. Explain the speed – Torque characteristics of PMBLDC motor.
30. With neat sketch, explain the constructional features of Permanent Magnet Synchronous Motor.

31. Describe the construction and performance of PMSM with neat diagram.
32. Explain the torque- speed characteristics of PMBL SNW or PMSM motor.
33. State the important features of permanent magnet synchronous motor.
34. What are the types of materials used in permanent magnet motor?
35. What are the types of PM synchronous motor?
36. List out the differences between the PM brushless DC motors and PM synchronous motors.
37. State the drawbacks and applications of PMSM.

Short answer Questions:

1. What is a synchronous reluctance motor?
2. What are the types of synchronous reluctance motor?
3. Mention some applications of synchronous reluctance motor.
4. Draw the torque-angle characteristics of synchronous reluctance motor.
5. State the principle of operation of synchronous reluctance motors.
6. What are the types of rotor available in synchronous reluctance motor?
7. Mention any two advantages of synchronous reluctance motors.

Numerical Problems

1. A synchronous reluctance motor has 8 poles and operates at 60 Hz. Determine the synchronous speed and the number of rotor teeth required to achieve a specific pole pitch.
2. In a switched reluctance motor with 6 stator poles and 8 rotor poles, calculate the step angle and the corresponding step resolution if the rotor has 24 teeth.
3. For a stepper motor with 200 steps per revolution and operating at 1000 rpm, calculate the time required for the motor to complete one full revolution.

PRACTICAL

1. Implement a closed-loop position control system using an AC or DC servomotor. Show how feedback from an encoder influences the motor's behaviour.
2. Demonstrate the use of servomotors in robotics by programming a robotic arm to follow a predefined path.
3. Showcase the precision positioning capabilities of stepper motors in applications like 3D printers or CNC machines.
4. Demonstrate the use of PMSMs in high-performance applications such as electric vehicles or industrial servo systems.
5. Implement closed-loop speed control for a BLDC motor using a microcontroller. Demonstrate the startup sequence and measure the motor's response to varying loads.
6. Show how BLDC motors are used in drone propulsion systems and demonstrate the advantages of their efficiency and compact design.
7. Showcase the suitability of SRMs in applications requiring high torque density and intermittent operation, such as in automotive systems.
8. Measure the synchronous speed of a SynRM and compare it with the theoretical calculation. Demonstrate how the rotor position affects the motor performance.

KNOW MORE

1. Special Electromechanical Systems (Introduction)
   https://nptel.ac.in/courses/108102156
2. Working of Synchronous Reluctance Motor
   https://en.engineering-solutions.ru/motorcontrol/syrm/
3. Working of Switched Reluctance Motor
   https://www.elprocus.com/switched-reluctance-motor-working
4. Working of BLDC
   https://www.elprocus.com/brushless-dc-motor-advantages-applications- and-control/
5. Working of Permanent Magnet Synchronous Motors
   https://www.linquip.com/blog/permanent-magnet-synchronous-motors/
6. Working of Stepper motors
   https://www.monolithicpower.com/en/stepper-motors-basics-types-uses
7. Working of AC and DC servomotors
   https://electricalworkbook.com/ac-servo-motor/

REFERENCES AND SUGGESTED READINGS

1. T.J.E.Miller,'Brushless Permanent-Magnet and Reluctance Motor Drives', Oxford University Press, 1989.
2. T. Kenjo, 'Stepping Motors and Their Microprocessor Controls', Clarendon Press London, 1984
3. K.Venkataratnam, 'Special Electrical Machines', Universities Press (India) Private Limited, 2008.
4. E.G. Janardanan, 'Special electrical machines', PHI learning Private Limited, Delhi, 2014.7.
5. R.Krishnan, 'Switched Reluctance Motor Drives – Modeling, Simulation, Analysis, Design and Application', CRC Press, New York, 2001.
6. T. Kenjo and S. Nagamori, 'Permanent Magnet and Brushless DC Motors', Clarendon Press, London, 1988.

Dynamic QR Code for Further Reading

APPENDICES

APPENDIX-A : Suggestive Template for Practicals

- Aim
  Explain briefly about the aim of the experiment.

- Relevance
  Explain about the relevancy of the experiment in your own words.

- Requirements
  List out all the required apparatus along with their proper specifications.

- Procedure, Observations and Inference
  Explain the procedure of the experiment step-wise and note the observations properly. On the basis of observations certain inference is to be made. You can use a table similar to that given below:

Step No.	Procedure	Observation	Inference
1			
2			
3			

- Video / animation
  If possible, you can go through some video/animation to visualize the steps physically.

- Calculations
  Properly calculate all the required physical quantities essential for your experiment.

- Result and Discussion (Error measurement)
  Obtain the final result and discuss about it with proper considerations of errors which can be introduced during your experiment.

- Conclusions
  Finally give your conclusion based on the obtained results.

- Validation of the topics in Experiment
  Try to validate the result of the experiment in real life scenario.

- Use of ICT
  You can also study using the available online resources. These are useful as there is no time constraint at all. Some of which are listed (not limited to) below:

  https://swayam.gov.in/
  https://nptel.ac.in/
  https://www.swayamprbha.gov.i/

APPENDIX-B : Indicative Evaluation Guidelines for Practicals / Projects / Activities in Group

Process Related Skills

Criteria and Level	Developing	Competent	Proficient
Handling the Set-up			
Recording of Data			
Time management			
Team Work			
Individual Work			
Safety Precautions			

Product Related Skills

Criteria and Level	Developing	Competent	Proficient
Content			
Research/Survey			
Use of latest Technology			
Stays on Topic			
Preparedness			
Confidence of Presentation			
ICT Usage including ppt Making Skill			
Time Management			
Group Efforts			
Individual Efforts			

APPENDIX-C : Assessments Aligned to Bloom's Level

- Bloom's Taxonomy – It has been coupled into following two categories for developmentof Questions for this Book as given below:

Category I Questions	Category II Questions *- Higher Order Thinking Skills*
Bloom's Level 1: Remember Bloom's Level 2: Understand Bloom's Level 3: Apply	Bloom's Level 4: Analyse Bloom's Level 5: Evaluate Bloom's Level 6: Create

APPENDIX-D : : Records for Practical

S. No.	Exp. No.	Name of the Experiment	Date			Marks	Signature
			Actual	Repeat	Remarks		
1.							
2.							
3.							
4.							
5.							
6.							
7.							

REFERENCES FOR FURTHER LEARNING

1. P.S. Bimbhra, Electric Machines, Khanna Book Publishing Co., New Delhi (ISBN: 978- 93- 6173- 294)
2. Mittle, V.N. and Mittle, Arvind., Basic Electrical Engineering, McGraw Hill Education New Delhi, ISBN :9780070593572
3. Kothari, D. P. and Nagrath, I. J., Electrical Machines, McGraw Hill Education. New Delhi, ISBN:9780070699670
4. Bhattacharya, S. K., Electrical Machines, McGraw Hill Education, New Delhi, ISBN:9789332902855
5. Theraja, B.L., Electrical Technology Vol-II (AC and DC machines), S.Chand and Co. Ltd., New Delhi, ISBN : 9788121924375
6. Sen, S. K., Special Purpose Electrical Machines, Khanna Publishers, New Delhi, ISBN: 9788174091529
7. Janardanan E. G, Special Electrical Machines, Prentice Hall India, New Delhi ISBN: 9788120348806
8. Hughes E., Electrical Technology, ELBS
9. Cotton H., Electrical Technology, ELBS
10. T.J.E. Miller,'Brushless Permanent-Magnet and Reluctance Motor Drives', Oxford University Press, 1989.
11. T. Kenjo, 'Stepping Motors and Their Microprocessor Controls', Clarendon Press London, 1984
12. K.Venkataratnam, 'Special Electrical Machines', Universities Press (India) Private Limited, 2008.
13. E.G. Janardanan, 'Special electrical machines', PHI learning Private Limited, Delhi, 2014.7.
14. R.Krishnan, 'Switched Reluctance Motor Drives – Modeling, Simulation, Analysis, Design and Application', CRC Press, New York, 2001.
15. T. Kenjo and S. Nagamori, 'Permanent Magnet and Brushless DC Motors', Clarendon Press, London, 1988.

CO AND PO ATTAINMENT TABLE

Course outcomes (COs) for this course can be mapped with the programme outcomes (POs) after the completion of the course and a correlation can be made for the attainment of POs to analyze the gap. After proper analysis of the gap in the attainment of POs necessary measures can be taken to overcome the gaps.

Table for CO and PO attainment

Course Outcomes	Expected Mapping with Programme Outcomes (1- Weak Correlation; 2- Medium correlation; 3- Strong Correlation)						
	PO-1	PO-2	PO-3	PO-4	PO-5	PO-6	PO-7
CO-1							
CO-2							
CO-3							
CO-4							
CO-5							

INDEX